Air Base Defense

Rethinking Army and Air Force Roles and Functions

ALAN J. VICK, SEAN M. ZEIGLER, JULIA BRACKUP, JOHN SPEED MEYERS

Prepared for the United States Air Force
Approved for public release; distribution unlimited

RAND PROJECT AIR FORCE

For more information on this publication, visit www.rand.org/t/RR4368

Library of Congress Cataloging-in-Publication Data is available for this publication.
ISBN: 978-1-9774-0500-5

Published by the RAND Corporation, Santa Monica, Calif.

© Copyright 2020 RAND Corporation

RAND® is a registered trademark.

Support RAND

Make a tax-deductible charitable contribution at
www.rand.org/giving/contribute

www.rand.org

Preface

The growing cruise and ballistic missile threat to U.S. Air Force bases in Europe has led Headquarters U.S. Air Forces Europe (USAFE) to reassess defensive options, both near and far term. In support of this reassessment, Headquarters USAFE asked RAND to explore the feasibility of USAFE acquiring ground-based missile defenses of its own and to consider the problem from both operational and service roles and functions perspectives. This report focuses primarily on issues relating to roles and functions, including the history of air base ground defenses, authorities for roles and missions, and case studies of Army–Air Force disputes regarding air base defense. It then assesses seven U.S. Air Force alternative courses of action to address air base active defense shortfalls. A companion volume discusses the operational aspects in greater detail.

The research reported here was commissioned by then–Brig Gen Charles Corcoran, Director of Operations, Strategic Deterrence and Nuclear Integration, Headquarters USAFE and Air Forces Africa, and conducted within the Strategy and Doctrine Program of RAND Project AIR FORCE as part of a fiscal year 2019 project, "Ground-Based Air and Missile Defense of Air Bases."

RAND Project AIR FORCE

RAND Project AIR FORCE (PAF), a division of the RAND Corporation, is the U.S. Air Force's federally funded research and development center for studies and analyses. PAF provides the Air Force with independent analyses of policy alternatives affecting the development, employment, combat readiness, and support of current and future air, space, and cyber forces. Research is conducted in four programs: Strategy and Doctrine; Force Modernization and Employment; Manpower, Personnel, and Training; and Resource Management. The research reported here was prepared under contract FA7014-16-D-1000.

Additional information about PAF is available on our website:
www.rand.org/paf/

This report documents work originally shared with the U.S. Air Force on May 31, 2019. The draft report, issued on September 11, 2019, was reviewed by formal peer reviewers and U.S. Air Force subject-matter experts.

Contents

Figures

Tables

Summary

The growing cruise and ballistic missile threat to U.S. Air Force bases in Europe has led Headquarters U.S. Air Forces Europe to reassess defensive options, including active ground-based systems that are currently assigned to the Army. The gap between the cruise missile threat and the U.S. joint force's capacity and capability to counter the threat is particularly worrisome. Constraints on resources and Army prioritization of mobile short-range air defenses for forward forces suggest that shortfalls in air base air defenses are likely to continue unless U.S. Department of Defense force planning and posture decisions give higher priority to these point defenses.

Approach

We broadly assessed threats, defense options, and constraints on roles and functions (R&F) to identify seven alternative courses of action for Air Force leaders to consider.[1] We then assessed the strengths and weaknesses of these prospective courses of action to determine whether they were likely to address the fundamental R&F issues. The historical R&F analysis at the core of this report uses previously unpublished primary source documents from the Air Force Historical Research Agency to gain new insights into the causes of past Army–Air Force disputes on air base defense.

We developed a framework to assesses how well service responsibilities for air defenses are aligned, comparing assigned responsibilities to service stakes, priorities, and force structures. The examples in Table S.1 illustrate two cases. In the first case, that of fleet air defense afloat, responsibilities, stakes, priorities, and force structure are well aligned. In the second case, that of ground-based air defense of Air Force bases, these factors are poorly aligned. Although the U.S. Army has been assigned the mission, that mission is not a priority for the Army, as reflected by the lack of dedicated forces. The Air Force is increasingly interested in acquiring its own ground-based air defense but does not believe it has the authority to do so.

---

[1] This analysis did not consider the acquisition, manning, or operations and maintenance costs that the Air Force would incur if it were assigned the air base air defense function. Such an analysis was outside the scope of this effort but would be natural follow-on research.

Table S.1. Examples of Well Aligned and Misaligned Service Responsibilities for Air Defense

	Example 1: Fleet Air Defense Afloat		Example 2: Ground-Based Air Defense of Air Force Bases	
	Navy	**Marines**	**Army**	**Air Force**
Service assigned responsibility?	Yes	Shared with Navy when afloat	Yes	No
Service with greatest stakes?	Yes	Shared with Navy when afloat	No	Yes
Service priority?	Yes	No	No	Growing
Dedicated force structure?	Yes	When afloat	No	No

Conclusions and Recommendations

We came to the following conclusions:

- Air base defense has been an enduring area of disagreement and frustration for the Army and Air Force.
- Although many factors are at play, the misalignment of service responsibilities and priorities for air base defense is hindering the correction of enduring shortfalls.
- The limitations of joint force development processes, Army resource constraints, and Air Force ambivalence have also contributed to an air base defense R&F roadblock.
- The Air Force may be able to bypass this roadblock through innovation and the use of advanced technologies, such as directed energy.
- The most robust strategy to improve air base defenses would pursue parallel lines of effort.

These led to the following recommendations:

- Demonstrate institutional commitment to air base defense by funding and advocating for substantial enhancements in capability areas already assigned to the Air Force, such as security forces and passive defense programs.
- Use the Air Force culture of innovation to break down R&F barriers.
- Propose a new memorandum of understanding with the Army to establish ground-based air defense of air bases as an Air Force responsibility.

Acknowledgments

The authors thank the project sponsors, Maj Gen Charles Corcoran and Brig Gen Michael Koscheski of U.S. Air Forces Europe A3 (USAFE/A3), and our point of contact, Max Hanessian (USAFE/A3Z), for their active involvement in the research and analysis. We also thank Karl Hebert (USAFE/A3Z) for his assistance in organizing a research trip to USAFE. Maj Charles Roper and Douglas Lomheim (both from USAFE/A3Z) graciously changed their travel plans to add a visit to the RAND Washington Office to brief the project team on threats and defensive options. Chris Rumley (Deputy Director, USAFE and U.S. Air Forces Africa History Office), John Williamson (Senior Historian, USAFE and U.S. Air Forces Africa History Office), William M. Butler (Air Combat Command Deputy Command Historian), Donald Fenton (Pacific Air Forces Command Historian), Daniel Haulman (Air Force Historical Research Agency [AFHRA]), Tammy Horton (AFHRA), Samuel Jackson (AFHRA), and Archie DiFante provided vital support and archival access for our research on roles and missions. We also thank Maj Gen John Wood (Commander, Third Air Force), COL David Shank (Commander, 10th Army Air and Missile Defense Command), Col Glen Christensen (U.S. Air Force A4S), Wing Commander Christopher Berryman (Royal Air Force Exchange Officer, U.S. Air Force A4S), and John Shackell (Air Force Security Forces Center) for sharing their views on air and missile defense of air bases. Col Christopher Corley (Chief of Security Forces, USAFE/A4S) met with us during our visit, and both he and his staff provided helpful feedback on the draft report.

Eric Pfister (U.S. European Command [EUCOM] J-5/8) kindly organized our EUCOM visit and acted as our escort. Finally, we thank staffs at EUCOM J3/5/8, EUCOM Joint Operations Center, NATO Headquarters Allied Air Command, and USAFE for their assistance with the research.

Thanks also to Joseph Shaw, Systems Planning and Analysis, Inc., for his constructive feedback on an earlier draft of the report.

At the RAND Corporation, we thank project members Daniel Norton and Russell Williams for their contributions. Thanks to RAND colleagues Paula Thornhill, Raphael Cohen, Miranda Priebe, David Ochmanek, and John Gordon for sharing their insights and expertise. Thanks to reviewers John Gordon, David Johnson, and Tom McNaugher for their thoughtful, expert, and constructive suggestions. Karin Suede and Silas Dustin provided administrative support to the research and helped prepare the manuscript.

Finally, we thank Phyllis Gilmore for her meticulous editing and exceptional efforts reviewing sources and citations for accuracy.

Abbreviations

AAA	antiaircraft artillery
AB	air base
ADC	air defense commander
AFB	Air Force base
AFH	Air Force handbook
AFTTP	Air Force tactics, techniques, and procedures
ALFA	Air-Land Forces Application Center
ALSA	Air Land Sea Application Center
AMD	air and missile defense
ARVN	Army of the Republic of Vietnam
AWACS	Airborne Warning and Control System
C2	command-and-control
CAS	close air support
CCD	camouflage, concealment, and deception
CEP	circular error probable
CHECO	Contemporary Historical Examination of Current Operations
CINCPAC	Commander-in-Chief, U.S. Pacific Command
CINCUSAFE	Commander-in-Chief, U.S. Air Forces in Europe
CIWS	Close-In Weapon System
CJCS	Chairman of the Joint Chiefs of Staff
COA	course of action
COIN	counterinsurgency
CONUS	continental United States
CORM	Commission on Roles and Missions
C-RAM	counter–rocket, artillery, and mortar system
CSA	Chief of Staff of the Army
CSAF	Chief of Staff of the Air Force
DALFA	Directorate of Air-Land Forces Application
DAS	Defense Acquisition System

DCA	defensive counter air
DJI	company name
DoD	U.S. Department of Defense
EA	electronic attack
EUCOM	U.S. European Command
EW	electronic warfare
FY	fiscal year
GAO	Government Accountability Office
GBAD	ground-based air defense
Gen 3	third generation
GNSS	global navigation satellite systems
GPS	Global Positioning System
HESCO	company name
HPM	high-power microwave
ICBM	intercontinental ballistic missile
IFPC	indirect fire protection capability
IRBM	intermediate-range ballistic missile
ISR	intelligence, surveillance, and reconnaissance
JCIDS	Joint Capability Integration and Development System
JCS	Joint Chiefs of Staff
JP	joint publication
JSA	joint service agreement
MACV	Military Assistance Command, Vietnam
MDA	Missile Defense Agency
MOB	main operating base
MOU	memorandum of understanding
MRZR	product designation
M-SHORAD	Maneuver Short-Range Air Defense
NATO	North Atlantic Treaty Organization
NASAMS	National Advanced Surface-to-Air-Missile System
NLOS	non–line of sight (missile)
NVA	North Vietnamese Army

OSD	Office of the Secretary of Defense
PACAF	Pacific Air Forces
PDF	portable document format
PPBE	Planning, Programming, Budgeting, and Execution System
R&D	research and development
R&F	roles and functions
RF	radio frequency
RAF	Royal Air Force
SAC	Strategic Air Command
SAM	surface-to-air missile
SHORAD	short-range air defense
SOF	special operations forces
SRBM	short-range ballistic missile
SUAS	small unmanned aerial system
THAAD	Terminal High-Altitude Area Defense
THOR	Tactical High-Power Microwave Operational Responder
UAS	unmanned aircraft system
UK	United Kingdom
USAAF	U.S. Army Air Forces
USAFE	U.S. Air Forces in Europe
USAREUR	U.S. Army Europe
USCINCEUR	U.S. Commander-in-Chief, European Command

1. Introduction

Background

The U.S. Air Force is dependent on airfields to conduct most of its core combat functions. Enemy attacks on runways, maintenance facilities, and fuel and munitions storage can inhibit or temporarily prevent the Air Force from generating combat power. Attacks on aircraft on the ground both limit sortie generation and attrite expensive and irreplaceable platforms. More broadly, reductions in intelligence, surveillance, and reconnaissance (ISR); air superiority; strike; and airlift sorties put the joint force at risk. Such reductions may also disrupt a theater campaign and will likely make a conflict longer and more costly.

Airmen and joint force leaders during World War II and the Cold War understood the necessity of air base defense well, but a period of rear-area sanctuary lasting from roughly 1990 to 2010 suggested that such lessons were no longer relevant.[1] While this sanguine view is no longer driving policy, two decades of counterinsurgency (COIN)–centric combat have diverted funding to other investment and operational priorities. The U.S. Department of Defense (DoD) and the services are now developing concepts and proposing programs appropriate for a fiercely contested future battlespace against a peer competitor, but change in large organizations is inherently slow and costly. Although the imperative of air base defense is often acknowledged, it must compete with other programs for funding and other fixed facilities for access to defensive systems.

The gap between the supply of short-range air and missile defenses, the demand for these systems, and past policy decisions regarding service responsibilities for short-range air defenses combine to yield a fraught policy environment and endangered bases. As it seeks enhancements to air base defense, the Air Force must find cost-effective solutions and, at the same time, either avoid capabilities that are assigned to other services or be granted the authority to develop such capabilities (either through negotiation or bureaucratic maneuvering).

The Policy Problem

The Air Force is the only military service that lacks clear authority to develop and procure surface-based air and missile defense (AMD) systems to protect its own forces.[2] The U.S. Navy

---

[1] For a discussion of the American way of war that developed during this sanctuary period, see Alan J. Vick, *Air Base Attacks and Defensive Counters: Historical Lessons and Future Challenges*, Santa Monica, Calif.: RAND Corporation, RR-968-AF, 2015a, especially Chapters Three and Four.

[2] By calling attention to this fact, we are not assuming that the Air Force should inherit such authorities or that doing so would automatically remedy any roles and functions (R&F) difficulties associated with defending air bases.

deploys SM-2, SM-3, RIM-116, and the Phalanx Close-In Weapon System (CIWS) to protect the fleet at sea from attack by aircraft and cruise or ballistic missiles. The U.S. Army deploys Terminal High-Altitude Area Defense (THAAD) and Patriot missiles to defend the theater from air and missile attack and counter–rocket, artillery, and mortar (C-RAM) systems to protect high-value fixed facilities from rocket, mortar, or artillery attack. It also deploys Avenger short-range air defense (SHORAD) systems to protect its maneuver forces and is developing new systems (Maneuver SHORAD [M-SHORAD] and the Indirect Fire Protection Capability [IFPC] program) to protect its forces and bases. The U.S. Marine Corps deploys Avenger to protect maneuver forces. The Air Force possesses fighter aircraft for air defense but has no ground-based defenses against attack by aircraft or missiles.

To be fair, the Air Force is not alone in this regard. As Figure 1.1 shows, 78 percent of the world's air forces lack ground-based air defenses (GBAD); 15 percent of air forces have surface-to-air missiles (SAMs); and only 7 percent possess both SAMs and antiaircraft artillery (AAA). That said, both the Russian and Chinese air forces deploy organic SAMs and AAA for air base defense.

Figure 1.1. Ground-Based Air Defenses in the World's Air Forces

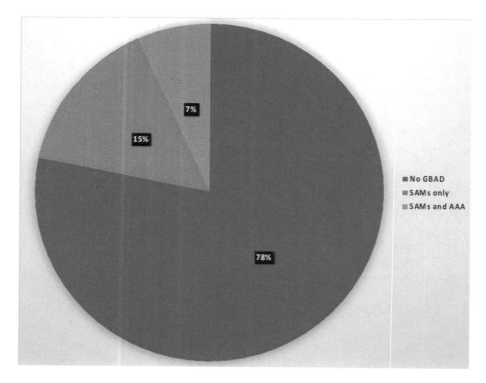

SOURCE: RAND analysis of 160 air forces in IHS Markit, *Jane's World Air Forces*, 2012.

Decisions made in the 1950s to divide point and area defenses between the Army and Air Force ultimately resulted in the Army owning all ground-based SAM and ballistic missile defense systems, whether for point or area defense. In principle, there is no reason that the Army cannot meet current Air Force requirements for air base defense. It did exactly this for selected Air Force bases between 1956 and 1966.[3]

During the Cold War, the Army dedicated an air defense artillery brigade to the defense of U.S. Air Forces Europe (USAFE) air bases.[4] But as Army SHORAD capabilities atrophied in the 1980s in response to the cancellation of the Sergeant York and Roland programs, the Air Force became increasingly concerned about the Army commitment to air base defense. As we will discuss in Chapter 5, this led to a 1984 memorandum of understanding (MOU) between the Army and Air Force that sought to clarify service responsibilities for this mission. Both services lost interest in the problem when the Cold War ended. Consequently, the Army retained the mission but devoted few resources to SHORAD systems of any type until recent geopolitical changes renewed interest in air defense.

The essential policy problem for the Air Force is how to ensure that air bases have effective defenses against air and missile attack, given institutional and budgetary constraints.

Research Approach

We used a variety of policy analysis methods to address this problem, including archival research based on primary source documents, a systematic review and integration of the scholarly literature, and textual analysis of key documents. The textual analysis was primarily qualitative, although we did create a simple program in Python to conduct keyword searches in Army *Air Defense Artillery* journal issues published between 1948 and 2006. Descriptive statistics were also created using data from public sources on the air defense capabilities of air forces around the world.

Finally, insights from technical and operational analyses were used to support the analysis of threats and defensive options presented in Chapters 2 and 3. This analysis did not consider the acquisition, manning, or operations and maintenance costs that the Air Force would incur if it were assigned the air base air defense function. Such an analysis was outside the scope of this effort. It would be valuable and natural follow-on research.

Purpose of This Report

This report is intended to help Headquarters USAFE, and the Air Force more broadly, assess the feasibility of deploying GBAD to defeat cruise missiles and other airborne threats. It considers the problem from both the operational and R&F perspectives.

---

[3] Mark Berhow, *U.S. Strategic and Defensive Missile Systems: 1950–2004*, Oxford, UK: Osprey Publishing, 2005, p. 62.

[4] Domenic P. Rocco, Jr., "Air Base Defense," *Air Defense Artillery*, Spring 1984.

Organization

Chapter 2 provides an overview of threats to air bases. Chapter 3 places active and passive defensive options in an R&F framework. Chapter 4 discusses the foundational documents and events that established postwar service R&F. It also presents the debates over air defense and specific developments regarding air base defense between 1948 and the end of the Vietnam War. Chapter 5 continues this historical analysis of R&F from Cold War Europe to today. Chapter 6 presents a framework to assess how well service responsibilities are aligned with other factors, especially service priorities, then identifies seven alternative Air Force courses of action (COAs) to address the misalignment of air base defense responsibilities. Chapter 7 presents our conclusions and recommendations.

2. Threats to Air Bases

Combatants of all types have long recognized air bases as worthy targets. The age of air warfare had barely begun when the Royal Navy Air Service successfully attacked the German Zeppelin base at Dusseldorf on October 8, 1914, destroying one of two Zeppelins based there.[1] Over the past 105 years of air warfare, air bases have been attacked on several thousand occasions in at least 26 separate conflicts. A diverse set of platforms and weapons has been used to attack air bases, including land- and sea-based fighter and bomber aircraft, unmanned aircraft systems (UASs), cruise missiles, naval gunfire, artillery, tank gunfire, rockets, and mortars. Additionally, commandos and terrorists have penetrated air base perimeters, using satchel charges, grenades, rocket-propelled grenades, small arms, and machine guns to destroy aircraft.[2]

This chapter offers a brief treatment of threats that USAFE bases might face over the next decade or so. The scale and severity of these threats vary greatly. For example, cruise and ballistic missiles are generally considered a much more serious threat to air bases than, for example, a lone UAS.[3] Nevertheless, in this chapter, we briefly consider the full range of potential threat weapons that employ either ballistic or aerodynamic principles, including civil aircraft and other possibilities that are less worrisome.

This more comprehensive approach has several benefits. First, by considering the full range of threats, we may identify hybrid threats that might otherwise be overlooked. Second, several current defensive systems have inherent potential against multiple threats. Because developing separate stovepiped defenses against each of the threat categories is unlikely to be affordable or operationally effective, taking a more holistic view of the threat can help ensure that future air base defenses are robust across a wider array of potential threat systems.

---

[1] See Ian Castle, *The Zeppelin Base Raids: Germany 1914*, Oxford, UK: Osprey Publishing, 2011, pp. 22–26.

[2] For an overview of air base attack tactics and aircraft losses by conflict, see Vick, 2015a. Other historical works on air base attack and defense include John Kreis, *Air Warfare and Air Base Air Defense, 1914–1973*, Washington, D.C.: Office of Air Force History, 1988; Roger P. Fox, *Air Base Defense in the Republic of Vietnam, 1961–1973*, Washington, D.C.: Office of Air Force History, 1979; Shannon W. Caudill, ed., *Defending Air Bases in an Age of Insurgency*, Maxwell Air Force Base (AFB): Air University Press, 2014; and Alan J. Vick, *Snakes in the Eagle's Nest: A History of Ground Attacks on Air Bases*, Santa Monica, Calif.: RAND Corporation, MR-553-AF, 1995.

[3] For detailed discussions of these threats, see Carl Rehberg and Mark Gunzinger, *Air and Missile Defense at a Crossroads: New Concepts and Technologies to Defend America's Overseas Bases*, Washington, D.C.: Center for Strategic and Budgetary Assessments, 2018; Defense Intelligence Ballistic Missile Analysis Committee, *2017 Ballistic and Cruise Missile Threat*, Wright-Patterson AFB, Ohio: NASIC Public Affairs Office, June 2017; and Eric Heginbotham, Michael Nixon, Forrest E. Morgan, Jacob L. Heim, Jeff Hagen, Sheng Li, Jeffrey Engstrom, Martin C. Libicki, Paul DeLuca, David A. Shlapak, David R. Frelinger, Burgess Laird, Kyle Brady, and Lyle J. Morris, *The U.S.-China Scorecard: Forces, Geography, and the Evolving Balance of Power, 1996–2017*, Santa Monica, Calif.: RAND Corporation, RR-392-AF, 2015, pp. 55–68.

Because the focus is on AMD, our discussion will be limited to threat weapons that current and prospective active defenses could reasonably be expected to defend against. Thus, we will consider missiles, rockets, aircraft, mortars, and UASs but not small arms and other weapons or munitions that ground forces might use if they penetrated a base perimeter. Defensive counters to threats will be discussed in the following chapter.

Cruise Missiles

Cruise missiles, a weapon that has proliferated widely, create difficult operational problems for air base defenders.[4] This section explains the characteristics of cruise missiles and the threat they pose to air bases.

A cruise missile is a pilotless, expendable airplane that harnesses aerodynamic lift to stay airborne until striking its target.[5] While the first operational cruise missile, Germany's V-1, was used during World War II to terrorize the British civilian population, modern cruise missiles have become effective weapons to accomplish a wide range of strategic and tactical missions, including precision attacks on fixed and semifixed land targets (e.g., air bases, long-range SAMs, and command posts) and attacks on ships at sea.[6] Modern cruise missiles are precise, difficult to detect, and affordable for many nations. These missiles can be fired from ground launchers, aircraft, ships, and submarines.[7]

The difficulty of detecting modern cruise missiles owes in part to their low radar cross section and their ability to fly earth-hugging flight profiles. Many cruise missiles have aerodynamic designs and other engineering features created specifically to reduce radar returns.[8]

The ability of cruise missiles to fly low-profile routes and to use terrain features to evade radar further compounds the defender's detection problem.[9] Ground-based radars struggle to detect low-flying cruise missiles because of terrain masking. Modern cruise missiles can also take circuitous routes to avoid radar detection, a further complication for defenders.[10] The speed of cruise missiles can also pose challenges: At supersonic speeds, these missiles' short flight

---

[4] For an early, but thorough, treatment of cruise missiles, see Richard K. Betts, ed., *Cruise Missiles: Technology, Strategy, Politics*, Washington, D.C.: Brookings Institution, 1981.

[5] Dennis M. Gormley, Andrew S. Erickson, and Jingdong Yuan, *A Low-Visibility Force Multiplier: Assessing China's Cruise Missile Ambitions*, Washington, D.C.: National Defense University Press, 2014, p. 2. Our definition adds *expendable* to emphasize that cruise missiles are a "one-time use" weapon.

[6] Kenneth P. Werrell, *The Evolution of the Cruise Missile*, Maxwell AFB, Ala.: Air University Press, 1985, p. 41.

[7] Gormley, Erickson, and Yuan, 2014; Defense Intelligence Ballistic Missile Analysis Committee, 2017, pp. 34–37.

[8] Low-observable cruise missiles are likely to proliferate in the future. As a result, more missiles will leak through forward and area defenses, reducing early warning for point air defenses and potentially increasing the number of missiles arriving in a wave. Future sensors, command-and-control (C2) networks, and terminal defenses will need to adapt to better detect and respond to such threats.

[9] Gormley, Erickson, and Yuan, 2014, p. 10.

[10] Defense Intelligence Ballistic Missile Analysis Committee, 2017, p. 35.

times reduce the time window available for the defender to react; airborne sensors may have difficulty distinguishing slow-moving cruise missiles from small civilian aircraft or ground clutter.[11] Unlike the German V-1, modern cruise missile systems are highly accurate and benefit from access to the Global Positioning System (GPS) or equivalent satellite technology.

Of note, cruise missiles are often categorized as land-attack cruise missiles or antiship cruise missiles. Either can be launched from land, sea, or air. For example, the Chinese CJ-20 is the air-launched version of the CJ-10 cruise missile (see Figure 2.1).[12] This conceptual division refers to the target of the cruise missiles—targets on land or ships at sea—and has implications for the engineering design of the missile. This section addresses land-attack cruise missiles, given our focus on the threat of cruise missiles to air bases.[13]

Figure 2.1. Chinese CJ-20 Air-Launched Cruise Missile

SOURCE: Wikimedia Commons.

The qualities just described make cruise missiles a serious threat to U.S. air bases—their planes, runways, fixed facilities, and personnel. We next review several operational analyses that

---

[11] John Stillion and David T. Orletsky, *Airbase Vulnerability to Conventional Cruise-Missile and Ballistic-Missile Attacks: Technology, Scenarios, and U.S. Air Force Responses*, Santa Monica, Calif.: RAND Corporation, MR-1028-AF, 1999, pp. 16–17.

[12] For more on the CJ-20, see Defense Intelligence Ballistic Missile Analysis Committee, 2017, pp. 34–37.

[13] For a discussion of antiship cruise missiles and their implications for military force planning, see Thomas G. Mahnken, *The Cruise Missile Challenge*, Washington, D.C.: Center for Strategic and Budgetary Assessments, 2005, pp. 9–19.

have examined the effects of cruise missile attacks on the operation of air bases and then discuss the general threat that cruise missiles pose to air bases.

The potential threat of cruise missiles to air bases was first systematically analyzed by RAND Corporation researchers John Stillion and David Orletsky in 1999.[14] Their analysis of attacks by cruise missiles armed with submunitions revealed that between 22 and 36 cruise missiles would achieve a 90-percent kill probability against aircraft parked in the open at four of the air bases the U.S. Air Force used during Operation Desert Storm.[15] A 2001 analysis by Dennis Gormley of a Chinese cruise missile attack on four Taiwanese air bases found that 75 missiles would achieve a 90-percent probability of closing the main runways and parallel taxiways.[16]

A 2015 RAND report analyzed the effect of Chinese cruise missile attacks on infrastructure targets at Kadena Air Base (AB) on Okinawa and Andersen AFB on Guam.[17] The analysis found that 60 cruise missiles could target every hangar, hardened aircraft shelter, and fuel tank at Kadena such that every target individually would suffer a greater than 90-percent probability of kill.[18] Similarly, 53 Chinese cruise missiles, including 33 with submunitions, could destroy aircraft parked in the open at Andersen and ensure a high probability of destroying the six hangars there.[19]

These analyses are not meant to suggest that cruise missile attacks are a silver bullet for U.S. adversaries or that they could be easily employed against U.S. air bases. To be maximally effective, cruise missile attacks require sophisticated integration with other attacks (e.g., ballistic missiles), real-time damage assessment, and a large cruise missile inventory—all exacting requirements. Additionally, the U.S. Air Force has adapted and will continue to adapt its operations to the cruise missile threat. Nonetheless, cruise missiles do constitute a significant threat to air bases, especially if used in salvos and in conjunction with other attacks. From the perspective of an air base defender, a massed cruise missile attack in which stealthy missiles attack from multiple directions and heights is a taxing scenario. Furthermore, the precision of cruise missiles makes them the most cost-effective weapon against hardened aircraft shelters, command posts, and other critical facilities, complementing attacks from ballistic missiles and special operations forces (SOF).

---

[14] Stillion and Orletsky, 1999.

[15] Stillion and Orletsky, 1999, p. 24.

[16] Dennis M. Gormley, *Dealing with the Threat of Cruise Missiles*, New York : Oxford University Press for the International Institute for Strategic Studies, 2001, p. 52.

[17] Heginbotham et al., 2015.

[18] Heginbotham et al., 2015, p. 63.

[19] Heginbotham et al., 2015, p. 63.

Ballistic Missiles

As with cruise missiles, the origins of ballistic missiles can be traced to the German military in World War II. Hitler's Germany fired approximately 3,000 V-2s, the first ballistic missile, at Britain and at other European countries.[20] Also as with early cruise missiles, the first ballistic missiles were inaccurate terror weapons, only able to hit large targets, such as cities. Recent developments, including improved accuracy and, especially, the addition of submunitions, have made ballistic missiles with conventional warheads a serious threat to U.S. air bases. This section discusses the basic characteristics of modern ballistic missiles armed with conventional warheads and then documents the threat this class of weapon poses for U.S. air bases in Europe and Asia.

Ballistic missiles—missiles initially powered by rockets that then follow a parabolic arc—have three phases of flight. The boost phase begins at launch and continues until the last rocket engine stops firing. The missile then enters a midcourse phase, in which it can travel at over 15,000 miles per hour, reducing the warning time available to any defender. Finally, during the terminal phase, the missile reenters the earth's atmosphere while still traveling faster than the speed of sound.[21] The high speed and altitude of ballistic missiles make interception extremely difficult and expensive. Ballistic missiles are often categorized by their range. See Table 2.1 for a definition of each category.

Table 2.1. Ballistic Missile Categories by Range

Missile Category	Range (km)
Short-range ballistic missile (SRBM)	>1,000
Medium-range ballistic missile	1,000–3,000
Intermediate-range ballistic missile (IRBM)	3,000–5,500
Intercontinental ballistic missile (ICBM)	>5,500

SOURCE: Center for Arms Control and Non-Proliferation, undated.

Ballistic missile payloads can also vary. Smaller SRBMs often carry payloads of approximately 500 kg; ICBMs sometimes carry payloads of several thousand kilograms.[22] Payload type also has important operational implications. Conventional ballistic missiles with a unitary warhead carry a single high-explosive charge. A ballistic missile armed with

---

[20] Christopher J. Bowie, *The Anti-Access Threat and Theater Air Bases*, Washington, D.C.: Center for Strategic and Budgetary Assessments, 2002, p. 37.

[21] Center for Arms Control and Non-Proliferation, "Ballistic vs. Cruise Missiles," fact sheet, undated.

[22] Among SRBMs, for instance, China's DF-11, DF-15, and DF-16 and Russia's SS-1, SS-21, and SS-26 all have payloads between approximately 500 and 1,000 kg. Russia's ICBMs, including the RS-24, SS-19, SS-25, and SS-27, all have payloads of at least 1,000 kg. China's ICBMs have similarly large payloads. See Missile Defense Project, "Missiles of the World," Missile Threat website, Center for Strategic and International Studies, undated.

submunitions can carry hundreds of small *bomblets*, miniaturized explosives that can be dispersed over a wide area.[23]

Finally, while ballistic missiles lack the pinpoint accuracy of cruise missiles, the accuracy of ballistic missiles, measured in circular error probable (CEP), has greatly improved in recent decades.[24] Some recent Chinese and Russian ballistic missiles are reported to have accuracies of approximately 50 m or less.[25]

The potential utility of nonnuclear IRBMs as airfield attack weapons was explored as early as the 1950s, but missile accuracies were deemed inadequate. A 1963 RAND report observed that, during the previous decade, the question of using ICBMs and IRBMs with nonnuclear warheads had "been posed many times and in most cases has been discarded."[26] Jaeger and Schaffer had determined that improving IRBM accuracies to 1,500-ft CEP would allow them to carry a sufficient number of submunitions (weighing 1.73 lbs each) to effectively attack aircraft parked in the open. However, the contemporary IRBMs had much larger CEPs—4,800 ft for Jupiter and 7,100 ft for Thor, for example—making the use of ballistic missiles to deliver nonnuclear ordnance on airfields infeasible.[27] By the mid-1960s, however, some analysts believed that Soviet IRBMs (e.g., the SS-4) could be modified to carry large payloads of submunitions against targets at distances under 1,000 km, trading range for payload. For example, a 1966 RAND report concluded that the SS-4 could achieve a CEP of 600 ft against targets within 1,000 km, with each missile delivering roughly 9,000 submunitions (weighing 0.63 lbs each).[28]

The Soviet Union did, in fact, deploy submunition warheads for its SRBM force in the following years, but, as late 1984, the intelligence community assessed that "the current SRBM nonnuclear threat to NATO air bases is marginal." The same assessment did note that "[w]hen the Soviets develop effective munitions to complement the projected terminal guidance (50-m CEP) capability of the improved SS-23, many of NATO's air defense aircraft could be pinned

---

[23] One RAND analysis of a potential submunition attack assessed that a Chinese-designed M-9 ballistic missile with a 500-kg payload could reasonably carry 825 bomblets. See Stillion and Orletsky, 1999, pp. 12, 14.

[24] CEP is the radius of a circle such that 50 percent of projectiles (missiles in this analysis) would land inside the circle.

[25] For a comprehensive listing of the CEPs of Chinese conventional ballistic missiles, see Jacob L. Heim, "The Iranian Missile Threat to Air Bases: A Distant Second to China's Conventional Deterrent," *Air and Space Power Journal*, Vol. 29, No. 4, 2015, p. 32. The Russian SS-16 Iskander is reported to have a CEP of 200 m or less depending on the particular guidance system employed. (Missile Defense Project, "SS-26 Iskander," Center for Strategic and International Studies Missile Threat website, December 19, 2019.)

[26] B. F. Jaeger and M. B. Schaffer, "Tentative Thoughts on Non-Nuclear IRBM's for Attacking Parked Aircraft," Santa Monica, Calif.: RAND Corporation, D(L)-11285-PR, May 17, 1963.

[27] Jaeger and Schaffer, 1963.

[28] John G. Hammer and W. R. Elswick, *Conventional Missile Attacks Against Aircraft on Airfields and Aircraft Carriers*, Santa Monica, Calif.: RAND Corporation, RM-4718-PR, 1966.

down for significant periods of time."[29] There is, however, no evidence that the Soviet Union deployed airfield attack SRBMs before the fall of the Berlin Wall in 1989.

With the end of the Cold War, the U.S. defense community largely lost interest in missile threats to air bases. A notable exception was the 1999 Stillion and Orletsky report.[30] They analyzed the ability of modern Chinese ballistic missiles armed with 1-lb submunitions to damage fighter aircraft parked in the open. The analysis indicated that one ballistic missile with conventional submunitions could achieve the same level of damage against aircraft parked in the open as eight ballistic missiles carrying unitary warheads.[31] Stillion and Orletsky also investigated the number of ballistic missiles with submunitions needed to achieve a 90-percent probability of kill against aircraft parked at four different postulated U.S. Air Force operating locations in the Middle East. Although their analysis examined different ballistic missiles and made a variety of assumptions about the lethal radius of the submunitions, the results suggested that, at most, 47 ballistic missiles were sufficient to achieve a 90-percent probability of kill.[32] Because China and Russia now each possess substantial conventional-capable ballistic missiles arsenals, the threat Stillion and Orletsky identified has only grown.[33]

Later analytical efforts also focused on the threat ballistic missiles posed to air base operations.[34] Recent modeling and simulation work at RAND has resulted in the creation of a suite of analytical tools able to evaluate the potential damage of ballistic missile attacks and potential counters to ballistic missiles.[35] Three recent open-source efforts stand out in their attempts to quantify the damage that ballistic missiles could inflict on U.S. air bases. *The U.S.-China Military Scorecard* effort examined the effect of ballistic missile salvos targeted at Kadena AB on Okinawa, Japan, and found that a large salvo could close the runway for days or even weeks.[36] Jacob Heim compared the relative effectiveness of Iranian and Chinese theater

---

[29] Director of Central Intelligence, *Warsaw Pact Nonnuclear Threat to NATO Air Bases in Central Europe*, Washington, D.C.: Central Intelligence Agency, NIE 11/20-6-84, October 25, 1984, p. 31.

[30] Stillion and Orletsky, 1999.

[31] Stillion and Orletsky, 1999, p. 12, footnote 5.

[32] Stillion and Orletsky, 1999.

[33] For China's conventional-capable ballistic missile arsenal size, see Heim, 2015, p. 32. For Russia's equivalent, see Billy Fabian, Mark Gunzinger, Jan van Tol, Jacob Cohn, and Gillian Evans, *Strengthening the Defense of NATO's Eastern Frontier*, Washington, D.C.: Center for Strategic and Budgetary Assessments, 2019, pp. 7–9.

[34] Bowie, 2002; Jeff Hagen, "Potential Effects of Chinese Aerospace Capabilities on U.S. Air Force Operations," testimony before the U.S.-China Economic and Security Review Commission, Santa Monica, Calif.: RAND Corporation, CT-347, May 20, 2010; Mark Gunzinger and Christopher Dougherty, *Outside-In: Operating from Range to Defeat Iran's Anti-Access and Area-Denial Threats*, Washington, D.C.: Center for Strategic and Budgetary Assessments, January 17, 2012.

[35] Brent Thomas, Mahyar A. Amouzegar, Rachel Costello, Robert A. Guffey, Andrew Karode, Christopher Lynch, Kristin F. Lynch, Ken Munson, Chad J. R. Ohlandt, Daniel M. Romano, Ricardo Sanchez, Robert S. Tripp, and Joseph V. Vesely, *Project Air Force Modeling Capabilities for Support of Combat Operations in Denied Environments*, RR-427-AF, Santa Monica, Calif.: RAND Corporation, 2015.

[36] Heginbotham et al., 2015, p. 61.

ballistic missiles against proximate air bases.[37] While arguing that the relative inaccuracy of their ballistic missiles complicates any Iranian attempt to shut down a U.S. air base in a potential conflict, Heim simultaneously demonstrated that Chinese ballistic missiles pose a substantial threat to U.S. air bases in Asia, especially given the limited basing options for the U.S. Air Force. Finally, Thomas Shugart and Javier Gonzales used two separate models to assess the Chinese threat to U.S. bases in Asia. The authors found that, no matter the model employed, "enough ballistic missiles seemed likely to leak through to cause highly significant damage to U.S. bases and forces in the region."[38] More-recent analyses have recognized the growing threat to air bases outside Asia, particularly the potential for Russian ballistic missile attacks on U.S. air bases in Europe.[39]

Finally, Iran's January 2020 attack on two air bases used by U.S. forces in Iraq (Al Asad AB and an airfield near Irbil) brings home the contemporary reality of this threat. In retaliation for the U.S. killing of Iranian Major General Qasem Soleimani, Iran launched more than a dozen ballistic missiles at the two locations, damaging and destroying some structures and causing traumatic brain injuries to 64 U.S. military personnel. This attack was relatively small and did not employ advanced submunitions, which was among the reasons that there were no deaths or damage to aircraft.[40]

Hypersonic Weapons

Recent testing of hypersonic missiles by Russia and China has generated intense interest in the threat these weapons pose and their potential contribution to U.S. long-range strike capabilities.[41] This section reviews the characteristics of hypersonic missiles and describes their threat to U.S. air bases.

---

[37] Heim, 2015.

[38] Thomas Shugart and Javier Gonzales, *First Strike: China's Missile Threat to U.S. Bases in Asia*, Washington, D.C.: Center for a New American Security, June 2017, p. 13.

[39] Fabian et al., 2019, pp. 7–9; Rehberg and Gunzinger, 2018, pp. 6–7.

[40] Neil Vigdor, "What We Know About the 2 Bases Iran Attacked," *New York Times*, January 7, 2020; Thomas Gibbons-Neff, "More American Troops Sustain Brain Injuries from Iran Missile Strike in Iraq," *New York Times*, January 30, 2020.

[41] For a representative sample of news coverage, see "What Are Hypersonic Weapons?" *The Economist*, January 3, 2019; Amanda Macias, "Russia and China are 'Aggressively Developing' Hypersonic Weapons," CNBC, March 21, 2018; and Geoff Brumfiel, "Nations Rush Ahead with Hypersonic Weapons amid Arms Race Fear," National Public Radio, October 23, 2018. For evidence of high-level U.S. government interest, see Thomas Karako, "Reenergizing the Missile Defense Enterprise," interview with Michael Griffin, Center for Strategic and International Studies website, December 11, 2018. For a comprehensive, although not up-to-date, report on hypersonic weapons, including Chinese and Russian programs, see James M. Acton, *Silver Bullet? Asking the Right Questions About Conventional Prompt Global Strike*, Washington, D.C.: Carnegie Endowment for International Peace, 2013, especially pp. 100–107.

Hypersonic weapons, defined as missiles that travel more than five times the speed of sound, can be subdivided into two types. Hypersonic cruise missiles are faster versions of traditional cruise missiles and, thus, experience powered flight throughout their trajectory. Hypersonic glide vehicles are typically launched on rockets into space, are released at high altitudes, and then "glide" to their targets along the upper atmosphere.[42] Both types of missiles pose a potential future threat to U.S. air bases.

Hypersonic missiles can fly at 5,000 to 25,000 kph and can have ranges over 10,000 km, enabling these missiles to hold targets at risk over broad regions.[43] Unlike ballistic missiles, however, these weapons do not follow a ballistic path and can alter their trajectories and associated impact points midflight. Additionally, hypersonic glide vehicles fly at lower altitudes than ballistic missiles, and hypersonic cruise missiles fly at higher altitudes than traditional cruise missiles.[44] Figure 2.2 compares the depressed trajectory of a hypersonic glide vehicle with the trajectory of a traditional ballistic missile. These qualities make missile defense, an already challenging mission, even more difficult.

Such high speeds compress the engagement timeline, leaving less time for targeting and assessment. Critically, they also reduce the time available for senior leader decisionmaking. The agility of hypersonic weapons also complicates the defender's task; the missile's ability to change course means that the impact point is highly uncertain until the final phase of flight. Maneuvering flight also makes interception difficult because ballistic models (that predict future location based on known characteristics of ballistic trajectories) cannot predict these in-flight changes in course.[45] Finally, the ability of hypersonic glide vehicles to fly at lower altitudes than ballistic missiles reduces the ability of land-based radars to detect and track these missiles; the ability of hypersonic cruise missiles to fly at higher altitudes than other cruise missiles allows these systems to fly above the altitude that most current SAM systems are capable of reaching.[46] In sum, hypersonic missiles are a formidable threat, and although no military has deployed a hypersonic missile yet, the active testing of these missiles makes deployment more likely.

---

[42] Richard H. Speier, George Nacouzi, Carrie A. Lee, and Richard M. Moore, *Hypersonic Missile Nonproliferation: Hindering the Spread of a New Class of Weapons*, Santa Monica, Calif.: RAND Corporation, RR-2137, 2017, pp. 2–3.

[43] By way of comparison, the distance between Moscow and Berlin is less than 2,000 km.

[44] Speier et al., 2017. For range information, see p. 9.

[45] This is not a new problem. Maneuvering reentry vehicles for ballistic missiles were tested in the United States as early as 1966, and extensive work continued through the 1970s. The only U.S. system to be equipped with a maneuvering reentry vehicle was the Pershing II IRBM. See Lauren Caston, Robert S. Leonard, Christopher A. Mouton, Chad J. R. Ohlandt, S. Craig Moore, Raymond E. Conley, and Glenn Buchan, *The Future of the U.S. Intercontinental Ballistic Missile Force*, Santa Monica, Calif.: RAND Corporation, MG-1210-AF, 2014, pp. 67–73.

[46] Speier et al., 2017, p. 12.

Figure 2.2. Hypersonic Glide Vehicle and Ballistic Missile Reentry Vehicle Trajectories and Detection Ranges by a Terrestrial Sensor

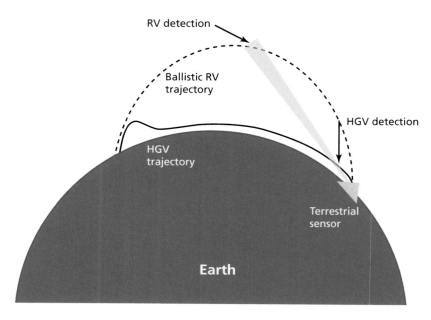

SOURCE: Speier et al., 2017, p. 11.
NOTES: Not to scale. HGV = hypersonic glide vehicle; RV = reentry vehicle.

If hypersonic missiles prove feasible and are procured in significant numbers, they could pose a serious threat to U.S. air bases in Europe and Asia. Russian and Chinese hypersonic missiles could reach all U.S. air bases in their respective theaters, including such bases as Andersen AFB in Guam or Royal Air Force (RAF) Lakenheath in England. The high speeds, maneuverability, and trajectories of these missiles will challenge U.S. and partner integrated AMD systems. Gen John Hyten, Commander of U.S. Strategic Command, has noted: "We don't have any defense that could deny the employment of [hypersonic missiles] against us."[47] General Hyten was referring to active defenses.[48] As we will discuss in the next chapter, passive defenses—such as camouflage, concealment, and deception (CCD), dispersal on and across bases, and hardening—all have the potential to reduce losses against hypersonic missiles and other threat weapon systems. That said, hypersonic weapons offer advanced adversaries another means of attacking aircraft in the open, facilities, personnel, and other targets on air bases.

---

[47] Barbara Starr, "US General Warns of Hypersonic Weapons Threat from Russia and China," CNN, March 27, 2018.

[48] John Grady, "DoD Official: U.S. Needs to Develop New Counters to Future Hypersonic Missiles," *U.S. Naval Institute News*, November 16, 2018.

Fixed-Wing Combat Aircraft, Civil Aircraft, and Larger Unmanned Aircraft Systems

In this section, we consider the threat posed by fixed-wing combat aircraft, light civil aircraft, and small UASs. Although joint and allied forward air defenses would likely stop most of these aircraft, some may penetrate and conduct reconnaissance against or directly attack air bases, especially those located in forward areas. In the case of general aviation and UASs, they might be launched from within the host nation, thereby avoiding forward air defenses. Thus, these threats represent potential targets for point air defenses.

Fixed-Wing Combat Aircraft

Fixed-wing aircraft threats to air bases could potentially include both bomber and fighter aircraft. Bombers could also seek to penetrate air defenses to deliver shorter-range weapons, but this would be unnecessary in many cases and, for most bombers, high risk. Therefore, bombers are most likely to act as platforms for standoff weapons, such as the Russian Kh-101 air-launched cruise missile.[49] For many missions, bombers can launch cruise missiles without breaching adversary air defenses.

In contrast, fighters typically must penetrate air defenses to deliver their weapons and generally are, at least somewhat, more survivable. Thus, air bases might come under attack by aircraft dropping laser- or GPS-guided munitions or dumb bombs. This threat is of greatest concern for forward air bases but could materialize as deep as the unrefueled combat radius of enemy tactical aircraft. U.S. planners certainly expected such attacks against NATO bases during the Cold War, assigning U.S. Army air defense batteries (typically a mix of Chaparral missiles and Vulcan guns) to the defense of individual air bases.[50]

Light Civil Aircraft and Unmanned Aircraft Systems

Air Force bases in Europe might also be subject to reconnaissance or attacks from light civil aircraft or longer-range UASs.[51] During a conflict, host-nation authorities will likely place

---

[49] Defense Intelligence Ballistic Missile Analysis Committee, 2017, p. 37; Douglas Barrie, "Kh-101 Missile Test Highlights Russian Bomber Firepower," Military Balance blog, February 8, 2019; and CSIS Missile Defense Project "Kh-101/Kh-102," Missile Threat webpage, June 15, 2018.

[50] For a contemporary assessment of the threat Warsaw Pact combat aircraft posed for NATO air bases, see Director of Central Intelligence, 1984, pp. 19–20. For details of U.S. European Command Cold War point-defense plans for air bases, see USAFE, *History of United States Air Forces in Europe for Fiscal Year 1972*, Vol. I: *Narrative*, Ramstein AFB, Germany: Office of the Command Historian, 1973, pp. 57, 59, and Rocco, 1984.

[51] For an assessment of UAV trends and policy implications, see Lynn E. Davis, Michael J. McNerney, James Chow, Thomas Hamilton, Sarah Harting, and Daniel Byman, *Armed and Dangerous? UAVs and U.S. Security*, Santa Monica, Calif.: RAND Corporation, RR-449-RC, 2014. Adversary helicopter gunships might also attack air bases, but, given their limited range and vulnerability to forward air defenses, this seems unlikely for all but the most forward air bases.

significant restrictions on general aviation flights originating within their borders. In a major war, many normal commercial flying activities would presumably be suspended. Nevertheless, it is possible that enemy agents or SOF might gain access to a civilian helicopter or light aircraft and use it for ISR or attack. Longer-range UASs might launch from enemy territory. For example, Russia possesses multiple UASs that can easily range across Europe from Kaliningrad or Russia proper, although their potential to penetrate forward air defenses seems very low.[52] Alternatively, larger UASs might launch from within host nations. This eventuality obviously presents a plethora of other problems for the attacker, including covertly shipping the UAS into the host nation and then launching it from an airstrip without detection. Wherever they are launched from, larger UASs could be used as ISR or weapon platforms.

While neither civil aircraft nor larger UASs are likely to appear in large numbers or be the preferred weapon delivery or ISR platform, a small number might be employed for high-leverage ISR, SOF missions, or strikes on high-value targets.[53] We do not consider these threats particularly serious, but a robust point air defense system should have some capability against such dangers.

Small Unmanned Aerial Systems

Small UASs (SUAS)—often defined as those weighing 55 lbs or less—have emerged as a distinct class of threat because of the rapid development of commercial SUAS technology, a series of high-profile international incidents involving everything from assassination attempts to airport closings, and even nascent battlefield use.[54] This section describes the military attributes of SUAS and their potential to disrupt U.S. air base operations.

SUAS possess two characteristics that make them more militarily useful than larger UASs or manned aircraft. First, the diminutive size, slow speed, and plastic construction of SUAS allows them to avoid detection by traditional antiaircraft sensors.[55] The systems' minute size and portability mean that even infantry and special forces can carry or transport SUAS on the

---

[52] For an assessment of the survivability of Predator-class UAS in contested air environments, see Joint Air Power Competence Centre, *Remotely Piloted Aircraft Systems in Contested Environments: A Vulnerability Analysis*, Kalkar, Germany, September 2014.

[53] See Ryan J. Wallace and Jon M. Loffi, "Examining Unmanned Aerial System Threats & Defenses: A Conceptual Analysis," *International Journal of Aviation, Aeronautics and Aerospace*, Vol. 2, No. 4, 2015; Davis et al., 2014; Brian A. Jackson, David R. Frelinger, Michael J. Lostumbo, and Robert W. Button, *Evaluating Novel Threats to the Homeland: Unmanned Aerial Vehicles and Cruise Missiles*, Santa Monica, Calif.: RAND Corporation, MG-626-DTRA, 2008.

[54] See Joint Publication (JP) 3-30, *Joint Air Operations*, Washington, D.C.: Joint Staff, July 25, 2019, p. III-31; Julia Macdonald, "The Most Surprising Thing About the Venezuela Drone Attack Is That It Hasn't Happened Sooner," Political Violence at a Glance website, September 4, 2018; Jon Gambrell, "How Yemen's Rebels Increasingly Deploy Drones," Defense News website, May 21, 2019.

[55] W. J. Hennigan, "Experts Say Drones Pose a National Security Threat—and We Aren't Ready," *Time*, May 31, 2018.

battlefield or behind enemy lines. Moreover, because SUAS are a relatively new phenomenon, defenses against them are similarly nascent. Second, because SUAS are relatively inexpensive, they can be procured and deployed in large numbers.[56] A swarm of SUAS, sharing information and coordinating their actions, represents the logical culmination of this military trend.[57] Swarming SUAS could disperse across an air base in search of the most lucrative targets, then, once one or more of the SUAS have detected the target(s), the remainder could swarm the target, producing mass effects well beyond what any single drone could effect.[58]

SUAS, however, have limited range, endurance and payloads. They generally offer shorter ranges than large UASs or manned aircraft do.[59] Additionally, the payloads of SUAS are necessarily limited, preventing any given SUAS from delivering a large weapon. That said, swarms of SUAS, each carrying a small munition, could attack a variety of targets individually or, in aggregate, swarm a target, possibly producing explosive effects equivalent to some larger unitary warheads. Figure 2.3 shows the widely proliferated DJI Phantom drone. This is the type of SUAS that might be used in a swarming attack.

Assuming sufficient range or clandestine employment, SUAS pose a serious threat to unprotected U.S. air bases. Equipped with explosives, SUAS could damage not only aircraft but also supporting equipment, such as fuel tanks and maintenance facilities or even exposed personnel.[60]

The prospect of SUAS swarms makes this threat all the more worrisome. In a 2017 interview with Defense News, Air Force Gen James M. "Mike" Holmes offered one possible scenario for swarming SUAS attacks on air bases: "Imagine a world where somebody flies a couple hundred of those down the intake of all my F-22s with just a small weapon on it?"[61] Even more concerning, SUAS swarms do not need to achieve technological maturity for SUAS to pose

---

[56] T. X. Hammes, "The Future of Warfare: Small, Many, Smart vs. Few & Exquisite?" War on the Rocks website, July 16, 2014; T. X. Hammes, "In an Era of Cheap Drones, U.S. Can't Afford Exquisite Weapons," Defense One website, January 19, 2016a; T. X. Hammes, "Technologies Converge and Power Diffuses: The Evolution of Small, Smart, and Cheap Weapons," Cato Institute website, January 27, 2016b; T. X. Hammes, "Cheap Technology Will Challenge U.S. Tactical Dominance," *Joint Force Quarterly*, Vol. 81, March 29, 2016c.

[57] John Arquilla and David Ronfeldt, *Swarming and the Future of Conflict,* Santa Monica, Calif.: RAND Corporation, DB-311-OSD, 2000; Sean Edwards, *Swarming and the Future of Warfare*, dissertation, Pardee RAND Graduate School, Santa Monica, Calif.: RAND Corporation, 2005; Paul Scharre, *Robotics on the Battlefield*, Part II: *The Coming Swarm*, Center for a New American Security, 2014; Paul Scharre, "Why You Shouldn't Fear 'Slaughterbots,'" *IEEE Spectrum*, December 22, 2017; Stuart Russell, Anthony Aguirre, Ariel Conn, and Max Tegmark, "Why You Should Fear 'Slaughterbots'"—A Response," *IEEE Spectrum*, January 23, 2018.

[58] For a discussion of swarming technology, see Amy McCullough, "The Looming Swarm," *Air Force Magazine*, March 22, 2019.

[59] Admittedly, some SUAS have flown long distances, even intercontinental distances, but most commercially available SUAS do not appear to be capable of flying long ranges. For an analytical perspective that emphasizes the long-range capabilities of some SUAS, see Hammes, 2014.

[60] See Hammes, 2016c.

[61] Valerie Insinna, "Small Drones Still Posing Big Problem for U.S. Air Force Bases," Defense News website, July 14, 2017.

problems for air base defense. In fact, the simple presence of large numbers of SUAS in the vicinity of an air base could halt or slow operations for fear of a U.S. aircraft colliding with a SUAS.[62] SUAS could also be used in a scouting role, providing intelligence, targeting, or battle damage assessment.[63]

Figure 2.3. DJI Phantom Drone

SOURCE: SachuHopes, 2018.

The threat of SUAS to air bases has, in fact, moved beyond the theoretical. Rebel groups in Syria have attacked a Russian air base with armed SUAS.[64] Rebels in Yemen have also attacked a civilian airport in Saudi Arabia, reportedly targeting a Patriot battery.[65] SUAS operating in civilian airspace have also led to airport closures, notably at Britain's Gatwick airport in December 2018.[66]

---

[62] Andrew Lacher, Jonathan Baron, Jonathan Rotner, and Michael Balazs, *Small Unmanned Aircraft: Characterizing the Threat*, McLean, Va.: MITRE Corporation, February 2019, p. 6.

[63] Scharre, 2014, pp. 27–30.

[64] David Reid, "A Swarm of Armed Drones Attacked a Russian Military Base in Syria," CNBC, January 11, 2018; Dmitry Kozlov and Sergei Grits, "Russia Says Drone Attacks on Its Bases in Syria are Increasing," Associated Press, August 17, 2018.

[65] "Yemen's Houthi Rebels Attack Saudi's Najran Airport—Again," Al Jazeera, May 23, 2019.

[66] Robert Wall, "U.K. Airport Remains Closed After Drones Disrupt Travel," *Wall Street Journal*, December 20, 2018; "Dubai Airport Grounds Flights Due to 'Drone Activity,'" BBC, September 28, 2016.

Rockets, Mortars, and Non–Line-of-Sight Missiles

Ground forces, including conventional forces, SOF, insurgents, and terrorists, have attacked air bases on at least 2,400 occasions since 1914.[67] The vast majority of the documented attacks occurred during Operation Iraqi Freedom or the Vietnam War, and the weapon of choice in both conflicts was the mortar or rocket.[68] Commando-style attacks, such as the January 5, 2020 attack by the Al-Shahab terrorist group against the U.S. airfield at Manda Bay, Kenya, are dramatic and can be quite destructive, but they are much harder to execute and relatively rare in modern conflicts.[69] These attacks feature ground forces penetrating the air base perimeter and destroying aircraft using small arms and explosives. Such attacks also fall outside the scope of this research. Our focus is on aerodynamic or ballistic weapons that are launched from outside the perimeter of an air base and, therefore, have the potential to be intercepted in flight.

For the purposes of this report, standoff attacks using rockets, mortars, or non–line-of-sight (NLOS) missiles are of primary interest because they all can be intercepted by ground-based defensive systems,[70] such as the U.S. Army's C-RAM system and the Israeli Iron Dome system—both of which likely have some capabilities against other airborne threats, such as

---

[67] The first documented ground attack on an airfield was during the German invasion of Belgium in World War I. On the evening of October 8, 1914, German forces began shelling the British airfield at Antwerp. Ironically, the first successful air attack on an enemy air base was launched from the same airfield earlier that day. See Castle, 2011, pp. 22–28.

[68] The total number of standoff attacks against U.S. air bases during operations Iraqi Freedom and Enduring Freedom is not publicly available. Data for attacks on Joint Base Balad between 2004 and 2010 are available, and the total is roughly 1,800. This suggests that the total for both conflicts could be well over 2,000. See Joseph A. Milner, "The Defense of Joint Base Balad: An Analysis," in Caudill, 2014, pp. 217–242. Viet Cong and North Vietnamese forces used mortars and rockets in 454 attacks against Air Force main operating bases (MOBs) in Vietnam. Penetrating attacks by "sapper" units accounted for another 21 attacks, for a total of 475. See Vick, 1995, p. 68.

[69] The British Special Air Service perfected such attacks during World War II, damaging or destroying almost 400 Axis aircraft in North Africa. Years later, during the Falklands War of 1982, the SAS conducted a similar raid against the Argentine airstrip on Pebble Island, destroying or damaging ten Argentine close support aircraft and one transport. See Vick, 1995, pp. 17, 37–65. More-recent commando attacks include the 2012 Taliban attack on Camp Bastion, Afghanistan, and the January 2020 Al-Shahab attack on Manda Bay. The 2012 Taliban attack destroyed six U.S. Marine Corps AV-8B Harriers and damaged another two. The 2020 Al-Shahab attack killed three Americans and either damaged or destroyed six surveillance and medical evacuation aircraft and a fuel storage tank. For details of the 2012 Taliban attack, see Alissa J. Rubin, "Audacious Raid on NATO Base Shows Taliban's Reach," *New York Times*, September 16, 2012. For details of the 2020 Shahab attack, see Thomas Gibbons-Neff, Eric Schmitt, Charlie Savage, and Helene Cooper, "Chaos as Militants Overran Airfield, Killing 3 Americans in Kenya," *New York Times*, January 22, 2020.

[70] An NLOS missile is an optically guided precision munition designed to strike point targets out to 20–30 km. It can attack moving targets, such as tanks, or static targets, such as parked aircraft. An example of such a missile is Rafael's Spike NLOS. See Yaakov Lappin, "Rafael Launches Spike NLOS from Tomcar Buggy," *Jane's Defence Weekly*, February 6, 2019.

UASs and, perhaps, cruise missiles.[71] Iron Dome, in particular, is seen as having potential against a wider range of threats.[72]

Rockets and Mortars

Rockets and mortars can be quite lethal against aircraft parked in the open. For example, during the Vietnam War, North Vietnamese Army (NVA) and Viet Cong rocket and mortar attacks on Air Force MOBs destroyed 94 aircraft and damaged another 1,149.[73] Between 1967 and 1968, losses to mortar and rocket attacks tripled despite an increase in the use of aircraft revetments. This led the Air Force to embark on an urgent program to construct hardened aircraft shelters. Close to 400 hardened aircraft shelters designed to defeat 122-mm rocket attacks were constructed between 1968 and 1970.[74]

Figure 2.4 illustrates the damage a relatively small mortar attack can do. This November 1964 attack was the first such attack on a U.S. Air Force base in Vietnam. Roughly 60 mortar rounds were fired from six 81-mm mortars by a company-size Viet Cong force during the 20-minute attack, destroying five U.S. Air Fore B-57s, heavily damaging another eight, and lightly damaging another seven aircraft, taking an entire squadron out of action.[75]

Mortars and rockets each have unique strengths and weaknesses. Man-portable rockets, such as the 122-mm rocket used during the Vietnam War, are simple weapons that can be fired in mass from basic launchers, such as bamboo poles or even shallow pits in the ground. Once they have fired, the attackers can escape without having to carry any heavy equipment. In contrast, mortars—consisting of a heavy base plate, tripod, tube, optics, and aiming mechanisms—are typically considered too valuable to be left behind. After the attack, the mortar must be dismantled and carried away. This may slow or complicate the attacking force's withdrawal. Thus, rockets have generally been the preferred choice for insurgent forces.

Rockets, however, lack accuracy; relatively large attacks are required to achieve a high probability of damage. Insurgents attacked U.S. air bases in both Iraq and Afghanistan with rockets but were largely unsuccessful due primarily to the small size of attacks and the effectiveness of the U.S. Army C-RAM system in shooting down incoming rockets.

---

[71] The U.S. Army C-RAM system mounts the Navy's Phalanx Gatling gun on an M916AE prime mover. The Army credits it with over 375 successful intercepts of rocket and mortar rounds fired at U.S. facilities. See U.S. Army, "Counter–Rocket, Artillery, Mortar (C-RAM) Intercept Land-Based Phalanx Weapon System (LPWS)," webpage, undated.

[72] Seth J. Frantzman, "Can Iron Dome Cut It for Indirect Fire Protection? U.S. Army Is Buying a Couple Systems to Find Out," Defense News website, February 6, 2019.

[73] At Air Force MOBs, 99 aircraft destroyed and 1,170 damaged in total. Five aircraft were destroyed, and another 21 were damaged by sappers (commandos) using grenades, satchel explosives, or small arms. Rockets and mortars were responsible for the rest of the losses. See Vick, 1995, pp. 68–89.

[74] Fox, 1979, p. 79; Karen Weitze, *Eglin Air Force Base: Installation Buildup for Research, Test and Evaluation and Training*, Eglin AFB: Air Force Materiel Command, 2001, pp. 239–240.

[75] For more on the November 1964 Bien Hoa attack, see Vick, 2015a, pp. 25–26.

Figure 2.4. B-57 Aircraft Destroyed During Mortar Attack, Bien Hoa AB, Vietnam, November 1, 1964

SOURCE: Photo courtesy of National Museum of the U.S. Air Force.

Mortar fire can be much more accurate than rockets but presents more-complex training and logistical challenges. It is also more difficult to mass mortar fire, although this issue is less critical because of the inherent accuracy of the weapon. To achieve high accuracy, the mortar must be emplaced carefully and, ideally, have its fire adjusted by a forward observer. Even without adjusted fire, the accuracy of mortars, combined with their natural dispersion patterns, makes them ideal weapons to use against area targets, such as troops in the open or aircraft parked on a ramp.[76] An example of a small man-portable mortar is the 60-mm weapon being fired by a U.S. Army Special Forces master sergeant in Figure 2.5. A more-capable but still man-portable mortar is the 81-mm used in the November 1964 attack described earlier.

---

[76] For additional details, see Vick 2015a, p. 28.

Figure 2.5. U.S. Army 60-mm Light Mortar

SOURCE: U.S. Army photo.

Mortar rounds are fired one at a time, with tube heating limiting the maximum rate of fire. Overheating can permanently damage the launch tube and also cause premature ignition of propellants. To avoid these problems, maximum and sustained rates of fire are established for a given mortar system. For example, the U.S. Army's 120-mm mortar can fire 16 rounds for 1 minute but then must stop firing to cool down. This system can, however, sustain a rate of four rounds per minute indefinitely.[77] Even a small mortar section (e.g., four mortars) could put considerable fire onto an aircraft parking ramp. For example, if four mortars each fire at the maximum rate of 16 rounds for 1 minute, 64 rounds could be sent downrange before counterfire could be placed on the mortar positions. This example reinforces the importance of C-RAM systems that can intercept mortar rounds in flight.

The Air Force is most likely to face mortar or rocket attacks in counterinsurgency (COIN) settings, but capable special forces might choose to use such attacks against Air Force bases. In

---

[77] FM 3-22.90, *Mortars*, Washington, D.C.: Headquarters, Department of the Army, December 2007, Table 5-1.

particular, a few mortars firing GPS-guided rounds could be highly effective against aircraft in the open, as well as personnel and other relatively soft targets.[78]

An NLOS missile offers attackers a relatively compact precision standoff weapon with a range of up to 30 km. The Rafael Spike NLOS is an example of such a weapon.[79] The missile is optically guided and would be particularly attractive for use against a high-value target. It is fired from a jeep-type vehicle. The main disadvantages of this system are its cost and integration with a military-type vehicle. Presumably, it could be modified to fire from a civilian vehicle. The system is less versatile than other options, and it would be difficult for SOF to employ more than a few in an air base attack. That said, this weapon does offer the means of precisely delivering a large shaped charge against a high-value target.

AeroVironment's Switchblade munition represents another standoff attack option. The weapon system is designed to provide infantry with a lightweight, disposable, beyond-line-of-sight, precision strike option against personnel or light vehicles. Armed with a grenade-class weapon, Switchblade is optically guided and can hit targets out to 10 km. The system weighs only 5.5 lbs, and several can easily be carried in a backpack. A few of these might be employed against a single high-value soft target, such as a large aircraft, or dozens might be used to attack aircraft, vehicles, and personnel on a flight line. If employed in a mass attack (i.e., dozens of munitions), Switchblade could be viewed as a precursor to the much larger swarming SUAS threat discussed in the previous section.[80]

Conclusions

Table 2.2 summarizes the capabilities of the threat attack vectors discussed in this chapter against selected airfield assets. The scoring is from the Air Force perspective: Green indicates little or no Air Force vulnerability to the attack vector; yellow indicates moderate Air Force vulnerability; and red indicates significant Air Force vulnerability. Airfield assets that might be attacked include aircraft parked in the open, fighter aircraft in closed Cold War–era third-generation (Gen 3) hardened aircraft shelters, aircraft in open-sided expeditionary-style hardened aircraft shelters, large maintenance hangars, operating surfaces (runways, taxiways, and parking ramps), fuel farms (collections of large fuel tanks), and personnel in the open.

---

[78] For more-detailed discussions of ground threats to air bases, see Vick, 2015a, pp. 24–32; David A. Shlapak and Alan Vick, *"Check Six Begins on the Ground": Responding to the Evolving Ground Threat to U.S. Air Force Bases*, Santa Monica, Calif.: RAND Corporation, MR-606-AF, 1995; and Vick, 1995.

[79] See Lappin, 2019. For more on the proliferation of precision weapons that might be used by SOF, terrorists, or insurgents, see James Bonomo, Giacomo Bergamo, David R. Frelinger, John Gordon IV, and Brian A. Jackson, *Stealing the Sword: Limiting Terrorist Use of Advanced Conventional Weapons*, Santa Monica, Calif.: RAND Corporation, MG-510-DHS, 2007.

[80] AeroVironment, "Switchblade," website, undated.

Table 2.2. Weapon Capabilities Against Airfield Assets

| | Aircraft in Open | Hardened Aircraft Shelters | | Hangars | Operating Surfaces | Fuel Farm | Personnel in Open |
		Gen 3	Expeditionary				
Swarming SUAS	Significant	Little	Significant	Moderate	Little	Little	Significant
Rockets and mortars	Significant	Little	Moderate	Moderate	Moderate	Moderate	Significant
Civil aircraft and larger UASs	Little	Little	Little	Little	Little	Little	Little
Combat aircraft	Moderate	Moderate	Moderate	Moderate	Moderate	Moderate	Moderate
Cruise missiles (w/unitary warhead)	Moderate	Significant	Significant	Significant	Moderate	Significant	Moderate
Ballistic missiles (with DPICM class submunition)	Significant	Little	Little	Moderate	Moderate	Moderate	Significant
Hypersonic weapons (with unitary warhead)	Moderate	Moderate	Moderate	Significant	Moderate	Moderate	Moderate

SOURCE: RAND analysis based on known or projected properties of various weapon types. See DoD 6055.9-STD, *DoD Ammunition and Explosives Safety Standards*, Washington, D.C.: Office of the Deputy Under Secretary of Defense (Installations and Environment), February 29, 2008, p. 335.
NOTES: DPICM = Dual-Purpose Improved Conventional Munition. The Gen 3 hardened aircraft shelter was built at Air Force bases in Europe and the Pacific during the Cold War. It has 18 in. of reinforced concrete on top of a corrugated steel liner, a 24-in. reinforced concrete rear wall, and a 12-in. concrete front door. Green = little Air Force vulnerability to attack vector, Yellow = moderate Air Force vulnerability, Red = significant Air Force vulnerability.

Cruise missiles armed with unitary warheads are poor choices to attack aircraft or personnel in the open or operating surfaces but would be highly effective against point targets, such as hardened aircraft shelters, hangars, and fuel tanks. Cruise missiles would also be a good choice to attack a single high-value aircraft in the open.

Ballistic missiles lack the accuracy of cruise missiles but often have larger payloads and, therefore, are an effective means of delivering large numbers of submunitions against area targets, such as aircraft and personnel in the open and operating surfaces. They are a poor choice for small, hardened point targets, such as hardened aircraft shelters.

Hypersonic weapons might be able to deliver submunitions, but this would likely require terminal maneuvers to slow the vehicle sufficiently to release the payload. We considered the capability of a hypersonic weapon armed with a unitary warhead. Such a weapon would have little capability against aircraft or personnel in the open, or operating surfaces. However, it would likely be accurate enough to strike a large target, such as a maintenance hangar. Whether the first generation of hypersonic weapons will have the accuracy to strike smaller targets, such as fuel tanks or hardened aircraft shelters for fighters, remains to be seen. These weapons have moderate capability against such targets.

Combat aircraft in sufficient numbers armed with unitary, submunition, and runway-busting munitions can wreak havoc on an airfield. A large attack of this type, however, seems unlikely, given NATO and U.S. air defenses. Thus, combat aircraft have moderate capability here, given the low probability that any single threat aircraft (let alone a large force) could reach an air base in the rear.

Civil aircraft and larger UASs are even less likely to pose a threat, and we have therefore assessed them as having little to no capability against the airfield assets. That said, an individual civil aircraft or UAS could attack a single high-value target, such as an aircraft in the open.

Swarming UASs are of greater concern because of their ability to share information among a large number of attackers and deliver small munitions against multiple soft targets. They are potentially effective against aircraft in the open, aircraft in open-sided hardened aircraft shelters, or against personnel. They pose little threat to hardened targets.

Finally, mortars and rockets, if GPS-guided, equipped with submunitions, or launched in large numbers, can be lethal against soft targets, such as personnel, vehicles, and aircraft in the open. Although these weapons cannot penetrate a Gen 3 hardened aircraft shelter, they might defeat temporary protective shelters. Rockets and mortars have also proven capable of igniting fuel storage tanks.[81]

In the next chapter, we consider the potential utility of active and passive defenses against these threats.

---

[81] For example, on April 13, 1966, a large mortar attack on the U.S. airfield at Tan Son Nhut destroyed a 420,000-gal. fuel storage tank. An Air Force report noted that the "flames soared hundreds of feet into the night sky." See Southeast Asia Team, Project CHECO, *Attack Against Tan Son Nhut: Project CHECO Southeast Asia Report*, Hickam AFB, Hawaii: Headquarters, Pacific Air Forces, 1966, pp. 1, 8.

3. Air Base Defense Options

Efforts to defend airfields date back to the early months of World War I, when the Germans introduced both active defenses (e.g., medium machine guns) and passive defenses (primarily CCD) to protect airfields. The state of the art improved greatly during World War II with the introduction of more-lethal active defenses, systematic hardening of bases, and sophisticated dispersal and CCD programs. During the Cold War, both NATO and the Warsaw Pact deployed air defenses around air bases, built hundreds of hardened aircraft shelters for fighters, constructed dispersal fields, and developed civil engineering techniques and capabilities to recover from enemy attacks. Although threat and defensive technologies have evolved considerably in the intervening decades, the fundamental air base defense concepts (e.g., point and area active defenses, hardening, dispersal) remain largely unchanged.[1]

In this chapter, we discuss the potential utility of active and passive approaches to defend against the threats presented in Chapter 2. Active defense options include radio frequency (RF) jamming, directed-energy weapons, guns, SAMs, and fighters flying defensive counterair (DCA) missions. Passive defense options include CCD, dispersing assets on a base, dispersing assets across multiple bases, hardening of facilities, and postattack recovery capabilities.[2] We first discuss each weapon system or approach individually and then assess their potential utility and versatility across the spectrum of threats.[3]

Electronic Warfare

Electronic warfare has become a DoD priority in recent years because of the rise of China and Russia as major security threats and their significant electronic warfare (EW) capabilities.[4]

---

[1] For a short historical overview of air base defensive options, see Vick, 2015a. The major historical work on air base defense is Kreis, 1988. A valuable resource on modern concepts and tactics for air base defense is Sal Sidoti, *Air Base Operability: A Study in Airbase Survivability and Post-Attack Recovery*, 2nd ed., Canberra: Aerospace Centre, 2001. The most comprehensive treatments of ground threats to air bases are Caudill, 2014, and Shannon Caudill, ed., *Defending Air Bases in an Age of Insurgency*, Vol. II, Maxwell AFB, Ala.: Air University Press, 2019.

[2] JP 3-01, Chapter V, offers doctrinal guidance for active and passive defense. See JP 3-01, *Countering Air and Missile Threats*, Washington, D.C.: Joint Staff, validated May 2, 2018.

[3] More-formal simulation and modeling of air base defense options is ongoing at RAND and elsewhere. For descriptions of two analytical approaches used to assess the proper mix of air base resiliency capabilities, see Jeff Hagen, Forrest E. Morgan, Jacob L. Heim, and Matthew Carroll, *The Foundations of Operational Resilience—Assessing the Ability to Operate in an Anti-Access/Area Denial (A2/AD) Environment: The Analytical Framework, Lexicon, and Characteristics of the Operational Resilience Analysis Model (ORAM)*, Santa Monica, Calif.: RAND Corporation, RR-1265-AF, 2016a; Thomas et al., 2015.

[4] One report from the Center for Strategic and Budgetary Assessments that makes the case that EW dominance is necessary for victory in future conflicts is Bryan Clark and Mark Gunzinger, *Winning the Airwaves: Regaining America's Dominance in the Electromagnetic Spectrum*, Washington, D.C.: Center for Strategic and Budgetary

Air Force doctrine defines *electronic warfare* as "military action involving the use of electromagnetic . . . and directed energy to control the EMS [electromagnetic spectrum] or attack the enemy."[5] There are three types of EW: electronic attack (EA), EW support, and electronic protection. Of the three, EA is the most salient for this discussion.

Air Force EA is understandably focused on increasing the survivability of friendly aircraft exposed to enemy radars and antiaircraft weapons (e.g., SAMs, air-to-air missiles, AAA). Interest and investment in EA capabilities has waxed and waned since the end of the Cold War but is now a priority for Air Force leadership,[6] as reflected in the new Electronic Warfare Roadmap.[7] The use of EA for air base defense has received relatively little attention, although this is changing now that counterdrone technologies are being actively pursued.[8]

For this report, we are interested in EA tactics and technologies (e.g., jamming and directed energy) that might be used to defeat enemy weapons or platforms conducting offensive operations against friendly air bases. As part of an integrated air base defense, EA capabilities in the form of directed-energy systems might disable or destroy incoming enemy weapon systems or delivery platforms. Alternatively, jamming could be used to confuse enemy navigation systems and sensors or disrupt communications, such as those between a commercial drone and its operator.

A historic example from World War II illustrates the use of EA against enemy navigation systems. The German Luftwaffe developed radio direction finding systems so that their bombers could more accurately navigate to targets in Great Britain. Once the British discovered that the Germans were using navigation beams, British scientists pursued various jamming and spoofing techniques to deny German bombers this navigation tool. In what became known as the "Battle of the Beams," the British sought not only to jam German signals with noise, but to warp the German signals so that the bombers would be led slightly off course, enough to miss their targets but not enough to be detected by observant navigators.[9]

Assessments, 2017. A related report analyzes historical battle network competitions to identify elements necessary for victory in modern warfare: John Stillion and Bryan Clark, *What It Takes to Win: Succeeding in 21st Century Battle Network Competitions*, Washington, D.C.: Center for Strategic and Budgetary Assessments, 2015.

[5] Annex 3-51, *Electronic Warfare*, Maxwell AFB: Curtis E. LeMay Center for Doctrine Development and Education, October 10, 2014; JP 3-13.1, *Electronic Warfare*, Washington, D.C.: Joint Staff, February 8, 2012.

[6] U.S. Air Force and NATO neglect of EW capabilities became apparent in 1999 during Operation Allied Force and were highlighted as a major lesson learned from that conflict. See Benjamin Lambeth, *NATO's Air War for Kosovo: A Strategic and Operational Assessment*, Santa Monica, Calif.: RAND Corporation, MR-1365-AF, 2001, pp. 114–116.

[7] Colin Clark, "Air Force Launches Electronic Warfare Roadmap: EMS ECCT 2.0," Breaking Defense website, April 24, 2019.

[8] Rachel S. Cohen, "The Drone Zappers," *Air Force Magazine*, March 22, 2019a.

[9] For a detailed discussion of the "Battle of the Beams," see Alfred Price, *Instruments of Darkness: The History of Electronic Warfare, 1939–1945*, Yorkshire, UK: Frontline Books, 2017, pp. 21–50. R. V. Jones, one of the British scientists at the heart of these efforts, also comments on them in his memoir: R. V. Jones, *Most Secret War*, London: Wordsworth Editions, 1978, pp. 92–100, 102, 127–128.

Electronic attack is a vast and technical topic that goes well beyond the scope of this analysis.[10] The details of cutting-edge EA capabilities are also not typically available in open sources. That said, EA represents an important dimension of a balanced air base defense and, thus, should be included, at least conceptually, in any assessment of air base defense options.

In the following subsections, we consider RF jamming and directed-energy weapons to illustrate the potential contribution of EA to air base defense.

Radio Frequency Jamming

Airmen have long used airborne EW systems to jam or spoof enemy radars and communications.[11] RF jamming also has a potential role in air base defense. In principle, any system that is dependent on RF links for control, navigation, or weapon fusing is potentially vulnerable to jamming. Targeting radars on aircraft and munitions are also potentially vulnerable to jamming. As mentioned earlier, RF jamming of enemy navigation systems dates back to World War II and might be effective against some modern navigation systems, particularly global navigation satellite systems (GNSS), such as the American GPS.

Jamming and spoofing of navigation signals as a form of EA is conceptually similar to British efforts in World War II to disrupt German navigation aids.[12] Surprisingly, GNSS jamming is seen more often in civil settings than in war zones and is routinely used by individuals and governments for a wide range of purposes. For example, in the United States, GPS jammers (available on eBay), although illegal, are being used by "truckers trying to avoid paying highway tolls, employees blocking their bosses from tracking their cars, [and] high school kids using them to fly drones in a restricted area."[13] Russian security forces reportedly use GNSS jamming to protect President Putin when he travels.[14] Finally, and most salient for this

---

[10] For a history of EW in U.S. air operations, see Alfred Price, *War in the Fourth Dimension: U.S. Electronic Warfare from the Vietnam War to the Present*, Mechanicsburg, Pa.: Stackpole Books, 2001. A useful, if somewhat dated, primer on EW tactics and systems is Doug Richardson, *An Illustrated Guide to the Techniques and Equipment of Electronic Warfare*, New York: Arco Publishing, 1985.

[11] For a technical overview of jamming techniques, see "EW Jamming Techniques," in Naval Air Warfare Center Weapons Division, *Electronic Warfare and Radar Systems Engineering Handbook*, Point Mugu, Calif.: Naval Air Warfare Center Weapons Division, October 2013, pp. 4-13 to 4-13.9.

[12] This vulnerability was recognized more than 20 years ago (Irving Lachow, "The GPS Dilemma: Balancing Military Risks and Economic Benefits," *International Security*, Vol. 20, No. 1, Summer 1995; Scott Pace, Gerald Frost, Irving Lachow, David Frelinger, Donna Fossum, Donald K. Wassem, and Monica Pinto, *The Global Positioning System: Assessing National Policies*, Santa Monica, Calif.: RAND Corporation, MR-614-OSTP, 1995).

[13] Kashmir Hill, "Jamming GPS Signals Is Illegal, Dangerous, Cheap and Easy," Gizmodo website, July 24, 2017. See also Jeff Coffed, *The Threat of GPS Jamming: The Risk to an Information Utility*, Melbourne, Fla.: Harris Corporation, 2016.

[14] Elias Groll, "Russia Is Tricking GPS to Protect Putin," *Foreign Policy*, April 3, 2019.

analysis, the U.S. military experienced GPS jamming of its systems during Operation Iraqi Freedom.[15] GPS jamming has also been an ongoing problem in recent operations in Syria.[16]

The previous chapter raised the possibility that air bases could be attacked by several GNSS-guided systems, including aircraft-delivered bombs, cruise missiles, ballistic missiles, UASs, and advanced mortar rounds.[17] Preventing these weapons from receiving GNSS signals could degrade their accuracy, although to what degree requires more detailed analysis.[18] One thing is clear: The U.S. defense community takes the GPS jamming and spoofing threat seriously and has increasingly moved toward dual-mode seekers and antijam features to counter this threat.[19] This suggests that jamming GNSS-guided systems in their terminal phases (e.g., from jammers located at air bases) is effective under some conditions. It is not clear whether GNSS-jamming against contemporary and emerging adversary military systems would be sufficiently effective or versatile to justify the required investments in equipment, personnel, training, and the like, but this is one possible role for EA in air base defense.

GNSS and RF jamming against commercial drones, on the other hand, is clearly worth pursuing for air base defense, at least as an interim solution. Their value is reflected in the number of counter-UAS systems that use one or both types of jamming. According to the Center for the Study of the Drone at Bard College, jamming the RF link between the drone and operator and/or jamming its GNSS link are the most common counter-UAS defenses, appearing in 96 different systems. Only 18 systems use the next-most-popular method, using nets to capture the drone.[20]

The Marine Air Defense Integrated System (Figure 3.1) is an example of a weapon that uses RF jamming to disrupt control links between the UAS and operator.[21] This appears to be

---

[15] A CENTCOM spokesman reported in 2003 that the United States had destroyed six Iraqi GPS jammers; see Anne Marie Squeo, "The Assault on Iraq: U.S. Bombs Iraqi GPS-Jamming Sites," *Wall Street Journal*, March 26, 2003.

[16] Courtney Kube, "Russia Has Figured out How to Jam U.S. Drones in Syria, Officials Say," NBC News, April 10, 2018.

[17] The potential that adversaries might one day use GPS-guided weapons against the United States was acknowledged in early cost-benefit assessments of GPS. On example is Lachow, 1995.

[18] Stillion and Orletsky may have been the first to propose GPS jamming for defense of air bases from cruise missiles; see Stillion and Orletsky, 1999, pp. 43–45.

[19] For an assessment that considers how GPS jamming affects weapon performance, see Jeff Hagen, David A. Blancett, Michael Bohnert, Shuo-Ju Chou, Amado Cordova, Thomas Hamilton, Alexander C. Hou, Sherrill Lingel, Colin Ludwig, Christopher Lynch, Muharrem Mane, Nicholas A. O'Donoughue, Daniel M. Norton, Ravi Rajan, and William Stanley, *Needs, Effectiveness, and Gap Assessment of Key A-10C Missions: An Overview of Findings*, Santa Monica, Calif.: RAND Corporation, RR-1724/1-AF, 2016b. For more on U.S. counters to GPS jamming, see Joe Gould, "Guided-Bomb Makers Anticipate GPS Jammers," Defense News website, May 31, 2015, and Andrew Liptak, "The U.S. Army Will Test a New GPS That's Resistant to Jamming This Fall," The Verge website, June 9, 2019a.

[20] Arthur Holland Michel, *Counter-Drone Systems*, Annandale-on-Hudson, N.Y.: Center for the Study of the Drone, February 2018, p. 5.

[21] Megan Eckstein, "Marines' Anti-Drone Defense System Moving Towards Testing, Fielding Decision by End of Year," USNI News website, March 11, 2019; Harry McNabb, "Invisible Interdiction: Air Force Awards Contract for

effective against current-generation commercial drones, but future autonomous drones or ones with jam-resistant communications may be less vulnerable to jamming.

**Figure 3.1. Light Marine Air Defense Integrated System Jammer
Mounted on Polaris MRZR Vehicle**

SOURCE: Photo courtesy of U.S. Marine Corps.

RF jamming might also be used against other threats, such as munitions that use proximity fuzes for targeting. For example, proximity fuzes use small radars to determine the distance from the target and to detonate prior to impact. This allows mortars, artillery, and air-delivered munitions to be air burst over ground targets (greatly expanding the zone of lethal effects beyond that of a ground burst) and air-to-air missiles to detonate near an aircraft (as opposed to requiring a direct hit), increasing the probability of damage. Proximity fuse jamming was pursued during World War II, and the Air Force held a patent for jamming air-to-air missiles until recently.[22] Modern munitions, however, possess several features designed to defeat proximity fuse jamming, suggesting that this technique may only be useful against old or crude systems.[23]

Rail-Mounted Anti-Drone System," *Drone Life*, June 12, 2019; Kyle Rempfer, "This Gun Shoots Drones out of the Sky," Defense News website, April 10, 2018.

[22] From 1969 to 2019, the Air Force held a patent for an airborne system that would jam the proximity fuze of enemy SAMs (Richard E. Marinaccio and Ward M. Meier, "Proximity Fuze Jammer," U.S. Patent Number US4121214A, December 16, 1969, via Google Patents website).

[23] A. Nasser, Fathy M. Ahmed, K. H. Moustafa, and Ayman Elshabrawy, "Recent Advancements in Proximity Fuzes Technology," *International Journal of Engineering Research & Technology*, Vol. 4, No. 4, April 2015.

Lasers

Solid-state lasers are attractive as future air base defense weapons for several reasons. They are accurate, can shift rapidly from one target to another, can be effective against a range of targets, and have a low cost per shot. Their disadvantages include limited range because of atmospheric conditions, the extended dwell time needed to heat the target enough to damage it, and high power and cooling requirements. The disadvantages are less problematic for short-range air base defense from fixed sites, where the systems can be larger and where the mobility requirements are modest. Exemplar targets for a laser defense system include UASs, cruise missiles, helicopters, civil aircraft, rockets, and mortars.[24] Ballistic missile warheads are designed to resist the extreme temperatures associated with reentry from orbit and are, therefore, a much more challenging target for lasers.[25]

In 2019, the U.S. Navy announced that the USS *Preble* will be the first *Arleigh-Burke*–class destroyer to be outfitted with the High Energy Laser and Integrated Optical-Dazzler with Surveillance system (known as HELIOS) 60-kW laser for close in defense against UASs and surface craft. The Navy hopes to move incrementally to more-powerful lasers able to defeat antiship cruise missiles.[26] This announcement followed the 2014 test of a 30-kW laser on the USS *Ponce*, an *Austin*-class amphibious warship, during a deployment to the Persian Gulf. The Navy had sufficient confidence in the system that the *Ponce*'s commander was authorized to use it as needed for ship defense against UAVs, helicopters, and attacking small boats.[27]

More salient for air base defenders, the Marine Corps sent a mobile, counterdrone laser weapon in the 2–10 kW class to the field for testing in 2019. The Compact Laser Weapon System will be mounted on a Joint Light Tactical Vehicle.[28] Similarly, the Army recently announced the selection of contractors to build a 50-kW laser mounted on a Stryker armored vehicle. This system is designed to shoot down UASs and is to be fielded in 2022. Both the Army and Marine Corps are likely to deploy lasers in the 10–50 kW range in the near term; these or similar Air Force–developed systems could become part of an integrated air base defense system. The greatest outstanding question is whether lasers in the 300–600 kW range are feasible for ground-based systems; 300 kW is generally considered the minimum for an effective cruise missile defense. Army technologists apparently believe 300-kW lasers are possible in the

---

[24] For an assessment of lasers in the air base defense role, see Rehberg and Gunzinger, 2018, pp. 17–25.

[25] Sydney J. Freedberg, Jr., "The Limits of Lasers: Missile Defense at Speed of Light," Breaking Defense website, May 30, 2014. See also Mark Gunzinger and Bryan Clark, *Winning the Salvo Competition: Rebalancing America's Air and Missile Defenses*, Washington, D.C.: Center for Strategic and Budgetary Assessments, 2016, pp. 43–44.

[26] William Cole, "USS Preble to Be First Destroyer Equipped with Laser Defense System," *Honolulu Star-Advertiser*, May 29, 2019.

[27] Sam LaGrone, "U.S. Navy Allowed to Use Persian Gulf Laser for Defense," USNI News website, December 10, 2014.

[28] Paul McLeary, "Marines Develop Laser to Fry Drones from JLTVs," Breaking Defense website, June 26, 2019a.

relatively near term, ending a program to build a 100-kW laser in favor of one that would create a laser in the "250-plus kW" range.[29]

High-Powered Microwave Systems

The Air Force Research Laboratory is exploring options to counter drone swarms, which might include dozens or even hundreds of small UASs. Systems designed to disable or to destroy small numbers of medium to large drones (e.g., guns or missiles) are not well suited to defeat large waves of very small drones. The laboratory's Tactical High Power Microwave Operational Responder (THOR) is designed to fill this capability gap (see Figure 3.2). THOR "uses short bursts of high-powered microwaves to disable" UASs at short ranges.[30] Another Air Force system (the Counter-Electronic High-Power Microwave Extended-Range Air Base Air Defense [CHIMERA]) is designed to defeat targets at medium to long ranges.[31] These systems appear to be viable near-term defenses against swarming drones and likely also have potential against cruise missiles and other airborne threats.[32]

Defensive Counterair

Airmen have long argued that air superiority is the foundation of airpower.[33] This was forcefully articulated by early airpower theorists Giulio Douhet and Gen William "Billy" Mitchell and most recently by retired Air Force Lt Gen David Deptula and Air Force Maj Gen Alex Grynkewich.[34] The essence of the argument was succinctly captured by the former Chief of Staff of the Air Force (CSAF), Gen Ronald Fogleman, and the former Secretary of the Air Force, Sheila Widnall, when they observed that air superiority gives U.S. forces "freedom from attack and freedom to attack."[35] As discussed in the previous chapter, U.S. forces can no longer count

---

[29] Sydney J. Freedberg, Jr., "New Army Laser Could Kill Cruise Missiles," Breaking Defense website, August 5, 2019b.

[30] Andrew Liptak, "The U.S. Air Force Has a New Weapon Called THOR That Can Take out Swarms of Drones," The Verge website, June 21, 2019b.

[31] Liptak, 2019b.

[32] Rehberg and Gunzinger, 2018, pp. 26–36.

[33] For historical assessments of the role of air superiority in war, see Benjamin Franklin Cooling, ed., *Case Studies in the Achievement of Air Superiority*, Washington, D.C.: Air Force History and Museums Program, 1994; and Richard P. Hallion, *Control of the Air: The Enduring Requirement*, Bolling AFB, D.C.: Air Force History and Museum Program, 1999.

[34] See Giulio Douhet, *The Command of the Air*, trans. Dino Ferrari, Washington, D.C.: Office of Air Force History, [1921, Italian] 1983; William Mitchell, *Our Air Force: The Keystone of National Defense*, New York: E. P. Dutton and Company, 1921; David Deptula, "America's Air Superiority Crisis," Breaking Defense website, July 12, 2017; and Alex Grynkewich, "An Operational Imperative: The Future of Air Superiority," *Mitchell Institute Policy Papers*, Vol. 7, July 2017.

[35] Ronald R. Fogleman and Sheila E. Widnall, *Global Engagement: A Vision of the 21st Century Air Force*, Washington, D.C.: Department of the Air Force, 1996.

on freedom from attack, especially from enemy cruise and ballistic missiles. Improvements in both offensive counterair and DCA capabilities can help mitigate the threat of attack from enemy aircraft, UASs, or missiles.[36]

Figure 3.2. The U.S. Air Force THOR High-Powered Microwave System

SOURCE: U.S. Air Force photo.

Current Air Force doctrine for control of the air encompasses both offensive and defensive operations. The counterair framework includes attacks on enemy air bases, missile sites, and related targets in offensive counterair. Air Force doctrine groups both active AMD and the full range of passive defenses under DCA.[37] For the purposes of this report, we will limit our discussion of DCA to the use of fighter or other aircraft for terminal defense of air bases against cruise missiles. As is typical in air base resiliency analyses, we treat passive defenses separately from DCA.

---

[36] One of the best primers on the counterair mission is James M. Holmes, *The Counterair Companion: A Short Guide to Air Superiority for Joint Force Commanders*, Maxwell AFB, Ala.: Air University Press, 1995.

[37] LeMay Center for Doctrine Development and Education, "Doctrine Advisory: Control of the Air," Maxwell AFB, Ala.: Air University, 2017.

In considering the contribution of airborne assets to DCA, we focused on airborne early warning and battle management aircraft (e.g., the E-3 Airborne Warning and Control System [AWACS]) and fighter aircraft (e.g., the F-15, F-16, F-22s, and F-35).[38] In a major war, these systems would likely fly surveillance and combat air patrols in forward areas, with fighters assigned to particular lanes.[39] These forward combat air patrols (along with Patriot missiles) could attrite enemy fighters, bombers (including those carrying cruise missiles) and cruise missiles launched from enemy territory (ground or air) or from under or on the sea. The forward air defenses are critical but cannot be expected to stop all threats, especially low-observable systems. Thus, an air defense in depth is preferred, backstopping the forward lanes with either area or point defenses. An example of the latter is the teaming of an airborne early warning and battle management platform with fighters to protect air bases from aircraft or cruise missile attack. The ideal air base cruise missile defense would include both ground-based (e.g., an IFPC-like system) and air elements. Given the current shortage of Army SHORAD systems available for point defense of fixed facilities and delays in Army procurement of IFPC, Air Force DCA assets may be called on to fill this mission. AWACS and fourth-generation fighters, such as the F-15 and F-16, offer a viable near-term option for this mission.[40]

In the future, other more exotic options may become viable. One possibility would be an "arsenal plane" concept in which a large aircraft with significant endurance and payload acted as both sensor and shooter. Alternatively, UASs might carry air-to-air missiles or high energy lasers for defense against UASs or cruise missiles.[41]

Short-Range Air Defense Systems

SHORAD systems provide the final active barrier to enemy attack. They complement forward air defenses, which, under ideal conditions, detect and attrite most of the attacking

---

[38] For more on the benefits of pairing airborne early warning and battle management platforms with fighter aircraft, see John Stillion, *Trends in Air-to-Air Combat: Implications for Future Air Superiority*, Washington, D.C.: Center for Strategic and Budgetary Assessments, 2015, pp. 27–28, 34. A riveting account of the fighter-AWACS team in action can be found in Dan Hampton, *Viper Pilot: A Memoir of Air Combat*, New York: William Morrow, 2012, pp. 67, 76–77, 79.

[39] For example, during Operation Desert Storm, "defensive fighter patrols were flown around the clock along the Saudi border" (Benjamin S. Lambeth, *The Winning of Air Supremacy in Operation Desert Storm*, Santa Monica, Calif.: RAND Corporation, P-7837, 1993, p. 4).

[40] Gunzinger and Clark, 2016.

[41] One version of the arsenal plane air-to-air platform is Eric Gons' proposal to use B-1s equipped with long-range AAM for forward defense in East Asia. For more on the B-1 idea, see Eric Stephen Gons, *Access Challenges and Implications for Airpower in the Western Pacific*, dissertation, Pardee RAND Graduate School, Santa Monica, Calif.: RAND Corporation, RGSD-267, 2011, pp. 133–151. Rehberg and Gunzinger, 2018, p. 17, propose arming UASs with high energy lasers for UAS and cruise missile defense.

force.[42] Earlier, we discussed the contribution EA systems can make to point defense. In this section, we consider more-traditional kinetic weapons, such as AAA and SAMs.

Antiaircraft Artillery

SHORAD systems consisted entirely of machine guns and cannons from World War I to the mid-1950s, with airfields among their many protected assets.[43] Because the most capable of the early weapons were artillery, these systems became known as AAA. In this subsection, we use *AAA* to refer exclusively to projectile-firing weapons, whether machine guns or cannon of various types. AAA rapidly proliferated and improved in lethality among all combatants during World War II when it was integrated with early warning radars, searchlight units, barrage balloons, and interceptor aircraft into air defense networks.[44] Even after the end of World War II, AAA units continued to provide the bulk of point air defense in the United States well into the 1950s, with 44 active-duty Army AAA battalions and 22 National Guard battalions deployed around cities, strategic air bases, and other vital installations.[45] As late as 1954, a U.S. Army AAA battalion continued to provide the only point air defense for the U.S. Air Force base at Sculthorpe, United Kingdom (UK), with 40-mm guns, a quad .50 caliber heavy machine gun, and smoke generators. However, the limitations of these systems—operating only in daylight and for targets below 3,000 ft—were increasingly recognized as unacceptable against the growing Soviet threat. Other U.S. Air Force bases in the UK were defended by equally obsolete British AAA units.[46] Similarly, some Strategic Air Command (SAC) bases in the United States were protected by Army AAA units (equipped with 75-mm guns) as late as 1956. Describing these units as "wholly deficient to cope with the threat," Gen Thomas D. White, Vice CSAF, called on the Army to remove them. Although the Army was rapidly moving to replace AAA units with SAMs, it nevertheless demurred, presumably to protect force structure.[47]

---

[42] A good primer on air defense systems and concepts is Mike B. Elsam, *Brassey's Air Power: Aircraft, Weapons Systems and Technology Series*, Vol. 7: *Air Defence,* London: Brassey's Defence Publishers, 1989.

[43] The first antiaircraft gun was actually designed much earlier. In 1870, Krupp, the German armament manufacturer, designed a 25-mm cart-based rifle to defeat French balloons. For a history of AAA, see Ian V. Hogg, *Anti-Aircraft Artillery*, Marlborough, UK: Crowood Press, 2002, pp. 10–11.

[44] A useful overview of World War II air defense concepts is found in Wesley F. Craven and James L. Cate, *The Army Air Forces in World War II,* Vol. VI: *Men and Planes,* Washington, D.C.: Office of Air Force History, [1955] 1983, pp. 78–118.

[45] James D. Crabtree, *On Air Defense*, Westport, Conn.: Praeger Publishers, 1994, p. 122.

[46] R. E. Tuck, *Preservation of Tactical Air Combat Potential in Western Europe: Guided Missile Defense Potential*, Santa Monica, Calif.: RAND Corporation, RM-1312, September 1954, pp. 148, 150, 161.

[47] L. H. Buss, Lloyd H. Cornett, Jr., Elsie L. Joerling, and Derril E. Howell, *Continental Air Defense Command Historical Summary: July 1956–June 1957*, Colorado Springs, Colo.: Continental Air Defense Command Office of History, September 15, 1957, p. 48.

Despite having largely been replaced by SAMS in forward and area defense roles, AAA has played a role in every major war to date; both the Russian and Chinese air forces retain AAA in their air defense units alongside their SAM systems.[48] As we will discuss in Chapter 5, the M163 Vulcan Air Defense System armed with a 20-mm gun (Figure 3.3) was one of two U.S. Army SHORAD systems (along with the Chaparral infrared [IR]–guided missile) in use through the first Gulf War. The cancellation of the Sergeant York gun (intended to replace the Vulcan) ultimately left the Army without an all-weather AAA system.

Figure 3.3. M163 Vulcan Air Defense System During Operation Desert Shield

SOURCE: U.S. Army photo.

As of early 2020, the only AAA type gun in the Army is the C-RAM system, which is based on the U.S. Navy's Phalanx system (essentially, a U.S. Navy version of the Vulcan).[49] C-RAM (Figure 3.4), however, is not intended for use against aircraft or cruise missiles. In contrast, the Navy continues to use the Phalanx CIWS for terminal defense of ships against a range of threats, including aircraft, antiship cruise missiles, UASs, and small vessels. CIWS is effective in this role because of the relatively small size of the defended target (a ship is tiny compared with an

---

[48] The information on Russian air force AAA comes from *IHS Jane's World Air Forces*, 2012. The Peoples' Liberation Army Air Force possesses both independent AAA regiments and brigades and integrated SAM/AAA units. See Bonny Lin and Cristina L. Garafola, *Training the People's Liberation Army Air Force Surface-to-Air Missile (SAM) Forces*, Santa Monica, Calif.: RAND Corporation, RR-1414-AF, 2016, p. 5.

[49] U.S. Army, undated.

air base), the typically clear fields of fire around a ship at sea, and the flight profiles of attacking weapons.

Figure 3.4. U.S. Army Counter–Rocket, Artillery and Mortar System

SOURCE: Photo courtesy of U.S. Army.

All calibers of rifles, machine guns, and artillery have been used in the antiaircraft role. The American .50 caliber and Soviet 12.7-mm heavy machines guns proved highly successful as multipurpose weapons that offered virtually any force, vessel, or installation (whether transportation truck company, merchant vessel, or encampment) some degree of self-protection against enemy aircraft. It was, however, the larger-caliber systems that proved most deadly and formed the foundation of integrated air defenses for many years. Rapid-fire 20-mm and 40-mm guns were complemented by much longer-range guns, such as the 75 mm, 88 mm, 90 mm, 105 mm, and 120 mm.

During World War II, German *Flugzeugabwehrkanonen* (aircraft defense cannon, famously abbreviated as *flak*) proved even more deadly than interceptor aircraft, downing 5,400 U.S. Army Air Forces (USAAF) aircraft, compared to 4,300 that Luftwaffe fighters downed, although only because the German high command invested vast sums equipping and manning this large force. Almost 2,000 flak batteries were assigned to Germany and the Western Front alone.[50] AAA proved quite deadly in more-recent wars as well. During the Vietnam War, gunfire was responsible for 89 percent of the losses of roughly 2,400 fixed-wing aircraft.[51] During the 1973 Yom Kippur War, a U.S. Army study estimated that the Israeli Air Force lost 31 aircraft to the ZSU-23-mm gun, 40 to radar-guided SAMs, and only four to the SA-7 man-portable air-defense system.[52]

Although AAA guns have been deadly in past conflicts and continue to pose a threat to aircraft flying at low altitudes, AAA fire is relatively ineffective for area defense against waves of small maneuvering targets, such as cruise missiles or SUAS. For that reason, specialized AAA units are not an attractive option for air base defense. The one possible exception is the C-RAM mission, which is currently accomplished using a gun system. But rapid advances in laser weapon technology suggest that the C-RAM mission will be one of the first to switch from guns to lasers.

On the other hand, advances in the integration of sensors, fire-control systems, and modern guns appear to give general-purpose weapons (e.g., 30-mm guns) a good self-defense capability. For example, a Stryker platoon equipped with the Army's Anti-Unmanned Systems Defense (AUSD) will be able to defend itself (or a small point target) from attack by an individual UAS or cruise missile.[53] In a similar vein, the Army and Air Force are experimenting with 40-mm rounds (containing nets) to defeat individual small drones. Reportedly, the system has proven accurate enough in tests to kinetically kill a drone without the net.[54] The Air Force Security Forces may want to pursue such systems because of their much greater lethality and versatility. But these systems would have to be deployed in large numbers to defend an air base from cruise missiles and UASs. In sum, guns are not the preferred means to defend an area target from waves of maneuvering cruise missiles or UASs.

Surface-to-Air Missiles

SAMs, both long- and short-range systems, have long been associated with air base defense. Short-range missile systems are receiving renewed attention because of their potential to defend against large volleys of cruise missiles.

---

[50] Edward B. Westermann, *Flak: German Anti-Aircraft Defenses, 1941–1945*, Lawrence, Kan.: University Press of Kansas, 2001, pp. 197, 286.

[51] Kenneth P. Werrell, *Archie, Flak, AAA, and SAM*, Maxwell, Ala.: Air University Press, 1988, p. 118.

[52] Werrell, 1988, p. 154.

[53] The AUSD integrates radar and EO sensors, a fire control computer, and the new Stryker 30-mm gun for the counter-SUAS mission. See Kris Osborn, "New Army Stryker 30mm Cannon Targeting Destroys Moving Drones," Warrior Maven website, April 29, 2019.

[54] Rachel Cohen, "Strategic Air Bases First to Receive Counter-UAS Systems," *Air Force Magazine*, July 1, 2019b.

The world's first SAM, the U.S. Army Nike Ajax (Figure 3.5) was deployed in 1953 to protect cities against Soviet bombers delivering nuclear gravity bombs. By 1956, four Nike batteries were also assigned to provide point air defense of SAC bases, initially at Fairchild AFB in Washington. As Table 3.1 shows, a total of 12 SAC bases in the United States and Greenland were eventually protected by either Nike Ajax or the more advanced Nike Hercules.[55]

Figure 3.5. U.S. Army Nike Ajax Missile

SOURCE: Photo by Sooe via Wikimedia Commons.

---

[55] An additional two bases had sites constructed but not occupied (Schilling and Walker AFBs), and five more were surveyed for sites that were not built: Columbus, Little Rock, Malmstrom, Mountain Home, and Sheppard AFBs (Berhow, 2005, p. 62).

Table 3.1. Nike Batteries Assigned to SAC Bases: 1957–1966

Base	Nike Ajax Batteries	Nike Hercules Batteries	Dates
Barksdale AFB, La.	0	2	1962–1966
Bergstrom AFB, Tex.	0	2	1960–1966
Dyess AFB, Tex.	0	2	1960–1966
Ellsworth AFB, S.D.	4	1	1957–1961
Fairchild AFB, Wash.	4	1	1956–1966
Lincoln AFB, Neb.	0	2	1960–1966
Loring AFB, Me.	4	2	1957–1966
Offutt AFB, Neb.	0	2	1960–1966
Robins AFB, Ga.	0	2	1960–1966
Thule AB, Greenland	0	4	1958–1965
Travis AFB, CA	4	2	1957–1971
Turner AFB, AL	0	2	1958–1965

SOURCE: Based on data in Berhow, 2005, p. 62.

Although the Nike Hercules proved to be useful in the strategic air defense role, it required the construction of expensive and elaborate facilities and, thus, had no ability to move with a modern Army or rapidly deploy to air bases in combat theaters. The Army began development of the Hawk missile system in 1952 to address the requirement for a smaller, deployable, and more mobile system with greater capability against low-flying aircraft. First deployed in 1960, the Hawk was hugely successful and was adapted for many different roles, including both forward barrier defense (in Europe) and point defense for air bases.[56] For example, as the U.S. military buildup in Vietnam began, commanders, fearing attacks from the North Vietnamese air force, requested that air defense units be deployed. Consequently, both U.S. Army and U.S. Marine Corps Hawk battalions were deployed. By the end of 1965, a Marine Corps Hawk battalion was in place to defend the Marine Corps–Air Force MOB at Da Nang, while Army Hawk units had deployed to defend Tan Son Nhut and Cam Ranh Bay air bases.[57] Hawks were primarily deployed in Europe as part of NATO's forward air defenses rather than being used for point defense. Twenty-five years later, Hawk and Patriot missile battalions deployed to Saudi Arabia as part of Operation Desert Shield, with King Abdul Aziz AB among the key installations protected.[58]

---

[56] Werrell, 1988, p. 92. For more on the Hawk missile system, see John A. Hamilton, *Blazing Skies: Air Defense Artillery on Fort Bliss, Texas, 1940–2009*, Fort Bliss, Texas: U.S. Army Air Defense Artillery Center, 2009, pp. 153–154.

[57] Wesley R. C. Melyan, *The War in Vietnam: 1965*, Hickam AFB, Hawaii: Headquarters, Pacific Air Forces, CHECO Division, January 25, 1967, p. 198.

[58] U.S. Air Force, *Gulf War Air Power Survey*, Vol. III: *Logistics and Support*, Washington, D.C.: Headquarters U.S. Air Force, 1993, p. 48, and Hamilton, 2009, pp. 294–301.

During the Cold War, the 108th Air Defense Artillery Brigade, the U.S. Army air defenders based in Germany, assigned a SHORAD battery for each protected airfield. A battery was composed of two Vulcan 30-mm gun platoons and two Chaparral missile platoons, fielding a total of eight Vulcan and eight Chaparral systems.[59] First deployed in 1970, the MIM-72 Chaparral (Figure 3.6) was a tracked SHORAD system equipped with the Sidewinder IR-guided missile. The gunner had to visually acquire the target and launch the missile using a visual sight. Although the Sidewinder was an effective missile, this limited Chaparral's use to daytime and clear weather. The Chaparral was replaced as the Army SHORAD system by the Avenger (see Figure 3.7), which included eight mounted Stinger IR-guided missiles and a .50 caliber machine gun. Like the Chaparral, the Avenger is a daylight, clear-weather system.[60]

Figure 3.6. MIM-72 Chaparral

SOURCE: U.S. Army photo.

---

[59] Rocco, 1984, p. 24.

[60] Hamilton, 2009, pp. 220–225, 287.

41

Figure 3.7. U.S. Army Avenger Air Defense System

SOURCE: U.S. Army photo.

Growing concern about Russian intentions and military capabilities has led to a revival of U.S. Army interest in air defense, especially mobile shorter-range systems necessary to protect maneuver forces. The Army is now pursuing both near-term measures, such as training Stinger teams, and new systems, such as the M-SHORAD, a concept that would integrate Stinger and Hellfire missiles and guns on a Stryker armored vehicle.[61]

Unfortunately, Army SHORAD systems face shortfalls in both capacity and capability. For example, the Army National Guard has only seven Avenger battalions, not nearly enough to defend Army maneuver forces and fixed facilities, such as air bases. Furthermore, the Avenger's Stinger missile is "not effective against modern cruise missile threats."[62]

The Army IFPC program seeks to address these shortfalls in counter–cruise missile and counter-UAS capabilities. IFPC is designed to provide 360-degree protection of fixed facilities. Based on the Sentinel radar, the Multi-Mission Launcher, and AIM-9XB2 missiles, the system has the potential to be highly effective against cruise missiles, either individually or in waves. Unfortunately, the IFPC program has experienced a variety of technical problems and appears to

---

[61] Randall McIntire, "The Return of Army Short-Range Air Defense in a Changing Environment," *Fires*, November–December 2017, p. 6; Gary Sheftick, "FY20 Budget to Boost Air and Missile Defense," U.S. Army website, March 13, 2019.

[62] Rehberg and Gunzinger, 2018, p. 12.

be a lower priority for the Army leadership than M-SHORAD and other air defense improvements needed for maneuver forces.[63]

Because of such delays, Congress took up the issue of cruise missile defense in its 2019 National Defense Authorization Act, singling out both the Secretary of Defense and the Army:

> If the Secretary of Defense certifies that there is a need for the Army to deploy an interim missile defense capability. . . , the Secretary of the Army shall deploy the capability as follows: (A) Two batteries of the capability shall be deployed by not later than September 30, 2020. (B) Two additional batteries of the capability shall be deployed by not later than September 30, 2023.[64]

The Army responded to the directive with plans to purchase two Israel-made Iron Dome short-range rocket defense system batteries, the only system that it can field quickly. Congressional legislation for the purchase of the systems, known as the U.S.-Israel Indirect Fires Protection Act, was introduced on June 10, 2019; the Army announced on August 8, 2019 that it had signed a contract with Raytheon and Rafael for delivery of the first two batteries in 2020.[65] It remains to be seen what capabilities this system, originally designed and used to intercept short-range rockets, offers in the counter–cruise missile role.

The failure of DoD to field capable defenses against modern cruise missiles is not the result of insurmountable technical challenges. At least one system that has already been fielded could be adapted for this role. The National Advanced Surface-to-Air-Missile System (NASAMS), developed jointly by the United States and Norway, uses the Sentinel radar and AIM-120 Surface-Launched Advanced Medium Range Air-to-Air Missiles and is "in service with seven countries and is part of the U.S. National Capital Region's air defenses."[66] NASAMS reportedly can engage 72 targets simultaneously.[67] The problem instead appears to be one of institutional priorities and enduring Army difficulties developing and acquiring complex systems.

---

[63] Reviewers Tom McNaugher and David Johnson both observed that the Army has often had problems developing and acquiring complex new systems. Both suggest that these acquisition process deficiencies are the driver of Army SHORAD shortfalls rather than a lack of commitment from the leadership. Regarding Cold War–era Army acquisition challenges, see Thomas L. McNaugher, *New Weapons, Old Politics: America's Military Procurement Muddle*, Washington, D.C.: The Brookings Institution, 1989.

[64] Pub. L. 115-232, The John S. McCain National Defense Authorization Act for Fiscal Year 2019, August 13, 2018, Section 112, pp. 1660–1661.

[65] U.S. House of Representatives, H.R. 3186, U.S.-Israel Indirect Fire Protection Act of 2019, 116th Cong., 2nd Sess., introduced June 10, 2019; Jen Judson, "It's Official: U.S. Army Inks Iron Dome Deal," Defense News website, March 26, 2019b.

[66] Rehberg and Gunzinger, 2018, p. 19.

[67] Missile Defense Advocacy Alliance, "National Advanced Surface-to-Air Missile System (NASAMS)," webpage, undated; Raytheon Company, "National Advanced Surface-to-Air Missile System," webpage, undated.

Passive Defenses

During World War II, airmen learned that air base design and operating practices (e.g., concentrating or dispersing aircraft) were vitally important.[68] When done poorly, these practices created vulnerabilities that adversaries could exploit, as demonstrated by the devastating Japanese attacks on U.S. airfields in Hawaii and the Philippines on December 7 and 8, 1941.[69] In contrast, a well-designed airfield complex (such as the RAF complex on Malta) that incorporated hardening, dispersal, camouflage, deception, and airfield recovery capabilities could sustain operations under heavy attack for long periods.[70] Even when an airfield had been poorly designed (e.g., the Japanese-constructed airstrip that the Marines captured and used on Guadalcanal), U.S. forces learned that dispersed operating concepts and airfield repair capabilities could enable airpower to survive and fight effectively even under heavy attack.[71]

Passive defenses include a variety of techniques that make it more difficult for adversaries to find friendly targets and, once found, to efficiently attack them. Passive defenses include hardening, dispersal of assets on a base, CCD, and distributed operations across bases.[72] In the following subsections, we briefly consider each of these options.

Hardening

Hardening refers to any actions taken to reduce the effects of enemy attack on airfield infrastructure and key resources (aircraft, personnel), including protection against kinetic, cyber, and EW weapons. Almost all air base hardening options require some specialized civil engineering and construction skills, such as those found in Air Force civil engineer squadrons. If personnel, equipment, and materials are available, expeditionary options can be constructed on the order of days to weeks. More-permanent options (e.g., Cold War–style hardened aircraft shelters) can take months to years to construct.[73]

---

[68] Perhaps the first systematic treatment of airfield design for resiliency is Merrill E. De Longe, *Modern Airfield Planning and Concealment*, New York: Pitman Publishing Company, 1943.

[69] Regarding the attacks on airfields in Hawaii, see Leatrice R. Arakaki and John R. Kuborn, *7 December 1941: The Air Force Story*, Hickam AFB, Hawaii: Pacific Air Forces Office of History, 1991. For more on airfield vulnerabilities in the Philippines, see William H. Bartsch, *December 8, 1941: MacArthur's Pearl Harbor*, College Station, Tex.: Texas A&M University Press, 2003.

[70] Kreis, 1988, pp. 111–136.

[71] For more on air operations out of Henderson Field and its dispersal strips during the most desperate fighting on Guadalcanal, see Thomas G. Miller, Jr., *The Cactus Air Force*, New York: Bantam Books, 1987; William Bradford Huie, *Can Do! The Story of the Seabees*, Annapolis, Md.: Naval Institute Press, 1997, pp. 39–44; and Eric M. Bergerud, *Fire in the Sky: The Air War in the South Pacific*, Boulder, Colo.: Westview Press, 2001, pp. 74–90.

[72] For an excellent overview of passive defense options, see Air Force Pamphlet 10-219, Vol. 2, *Civil Engineer Disaster and Attack Preparations*, Washington, D.C.: Department of the Air Force, June 9, 2008, pp. 27–54.

[73] During the Vietnam War, Air Force civil engineers built 373 hardened aircraft shelters in just 16 months (October 1968 to January 1970), an impressive rate. These shelters, however, were simple affairs by today's standards. The Vietnam shelters had the same overhead protection as a Cold War hardened aircraft shelter (a

The following discussion will, however, be limited to hardening against kinetic attack. Modest efforts to protect air bases via hardening (e.g., dugouts for personnel) began during World War I, but more-systematic hardening of airfields was not common until World War II.[74]

Hardening options typically include going underground (e.g., burying fuel storage tanks), adding protective barriers (e.g., revetments, sandbags, or HESCO bastions), or strengthening aboveground structures with reinforced concrete (e.g., hardened aircraft shelters).[75]

Going Underground

Going underground, either by excavating a hole or digging a tunnel into a mountainside, offers the greatest protective potential against kinetic attack. Underground structures range from the relatively shallow infantry-type dugouts (often dug by hand and covered with two-by-fours, plywood, and sandbags) used to protect airfield personnel during World War I and World War II to vast underground facilities, such as the alternate command center in the Cheyenne Mountain Complex that the North American Air Defense Command and U.S. Northern Command share.[76]

Some airfields in Sweden, Serbia, China, and Taiwan were built near large mountains so that aircraft could be sheltered in tunnels, but this option is limited to a small number of locations with permissible geography and geology—not particularly relevant for an expeditionary air force, such as the U.S. Air Force. At least one RAND report, in 1961, considered building underground shelters for SAC bombers—accessed via elevators—but the Air Force never pursued this, presumably because the expected protective benefit did not justify the cost and complexity.[77] Although the literature contains references to underground shelters for aircraft, we know of no case in which elevator-accessed shelters were constructed; *underground* always turns out to mean that the shelter was dug horizontally into a mountainside.

The underground options most salient for air base defense include bunkers for command posts or fuel storage. In general, these are constructed by first excavating a hole, then building some kind of hardened structure inside the hole, then covering that with soil and concrete. The depth and level of hardening varies as a function of the expected threat but need not be

---

corrugated steel arch covered with 18 in. of concrete) but had no front door and only a partial rear wall. See Fox, 1979, pp. 70–71.

[74] For an overview of airfield hardening in past conflicts, see Vick, 2015a, pp. 43–53. Also see Kreis, 1988, and Sidoti, 2001.

[75] HESCO is a British company that manufactures protective flood and security barriers. The barriers consist of foldable wire and fabric containers that can be erected quickly and filled (usually using a front loader) with soil, sand, or gravel. They come in various sizes and can be stacked to create a barrier that is roughly 10 ft tall and 3 to 4 ft thick. They provide good protection from small arms fire and mortar explosions. HESCO barriers became ubiquitous at U.S. installations during operations Enduring Freedom and Iraqi Freedom. See HESCO, "Mil Units: Protecting Forces Since 1991," webpage, undated.

[76] North American Aerospace Defense Command, "Cheyenne Mountain Complex," webpage, undated.

[77] John G. Hammer and Charles A. Sandoval, *Comparison and Evaluation of Protective Alert Shelters for SAC Aircraft*, Santa Monica, Calif.: RAND Corporation, D-8740, 1961.

particularly deep to provide protection from general-purpose surface-detonating weapons. Although such facilities exist at some U.S. Air Force bases, particularly in Europe and Korea, their number and size are typically limited because of cost and maintenance issues. In some cases, there may be cheaper alternatives. For example, simple slit trenches with overhead cover may be sufficient to provide easily accessible shelter for maintenance personnel on flight lines.

Adding Protective Barriers

Protective barriers of soil, brick, rock, concrete, wood, and steel are used individually or in combination to create berms or protective walls for aircraft or personnel.[78] During World War II, aircraft revetments were routinely used to provide protection from strafing and bombing.

Revetments proved valuable in limiting damage from bombing near misses and nearby secondary explosions and hindered strafing runs by enemy aircraft. Revetments could be constructed with relative ease by civil engineers and offered protection from both enemy attack and accidents with munitions or fuel. Revetments around a "victim" aircraft help contain explosions, fires, and fuel leakage (at least somewhat), while revetments around nearby aircraft help protect them from the results of attacks or accidents involving nearby aircraft. For these reasons, revetments have been (and are) used extensively by air forces around the world. That said, revetments offer no protection from a direct hit or a strafing run oriented toward the opening. These limitations led to the creation of hardened aircraft shelters, which we discuss in the next subsection.

Protective barriers are also used to protect personnel and structures from incoming mortars, rockets, bombs, and direct fire weapons. With overhead protection (e.g., a slab of concrete), an aboveground structure can provide good protection from mortar or rocket attack. Both concrete and HESCO bastion–type materials were used at air bases in Iraq and Afghanistan to provide personnel protection during mortar and rocket attack. These shelters were relatively small, typically housing a dozen or fewer individuals, but proliferated around bases so that shelters were easy to reach from sleeping quarters, dining facilities, and workspaces.[79]

---

[78] For an authoritative treatment of design and engineering aspects of revetment construction, see Air Force Handbook (AFH)10-222, Vol. 14, *Civil Engineer Guide to Fighting Positions, Shelters, Obstacles and Revetments*, Washington, D.C.: Department of the Air Force, August 1, 2008, pp. 108–128. Also see "Fortifications for Parked Army Aircraft," in FM 5-430-0-2 and Air Force Pamphlett 32-8013, Vol. II, *Planning and Design of Roads, Airfields, and Heliports in the Theater of Operations—Airfield and Heliport Designs*, Washington, D.C.: Headquarters, Department of the Army and Department of the Air Force, September 1994, Ch. 14.

[79] For an example of an aboveground personnel protective structure design using concrete slabs, see AFH 10-222, Air Force Tactics, Techniques, and Procedures (AFTTP) 3-32.34V3, *Civil Engineer Expeditionary Force Protection*, March 1, 2016, pp. 56–57; for discussion of other personnel protective structure designs, see AFH 10-222, pp. 62–71.

Strengthening Structures

As just indicated, revetments, although valuable, were far from perfect protection from air attack. As a result, reinforced concrete shelters were created to provide additional protection. The first hardened aircraft shelters were built by the Marine Corps at Ewa Field, Oahu. Marine aircraft on the ground at Ewa suffered heavy damage from the Japanese attacks on December 7, 1941. In response to this vulnerability, the Marines began building clamshell-type concrete shelters for their tactical aircraft in 1942. The shelters had 12 to 18 in. of reinforced concrete covered with soil. These shelters were open in front but otherwise provided excellent protection from strafing or bombing. These were the only hardened aircraft shelters the U.S. military built during World War II. As the Marines island-hopped across the Pacific, they never stayed anywhere long enough to justify such construction, although the Marines, Navy, and USAAF all made extensive use of revetments throughout the war.[80]

Hardened aircraft shelters made their next appearance during the Vietnam War. Revetments were widely used to provide some protection from mortar and rocket attack, but after 500 U.S. Air Force aircraft were damaged or destroyed by indirect fire attacks in 1968, Air Force leaders decided to take urgent action.[81] This resulted in a two-year program of shelter construction, initially just putting roofs on existing revetments. These shelters were made out of corrugated metal arches covered with 18 in. of concrete, sufficient to defeat up to 140-mm rocket attacks.[82] Beginning in 1969, similar shelters (now with hardened doors) were built at USAFE bases. Several iterations of designs eventually produced the hardened aircraft shelters found at Ramstein and other air bases in Europe.[83]

The fighter hardened aircraft shelters found at U.S. Air Force bases in Europe and Asia were well designed to defeat near misses from aircraft delivering dumb bombs and tactical missiles—the primary Cold War–era threats. These shelters were, however, never intended to defeat direct hits from precision weapons, such as modern cruise missiles and laser or GNSS-guided bombs. Nevertheless, these older shelters have value, offering good protection from near misses; small submunitions; mortars; rockets; UASs; and direct-fire weapons, such as rocket-propelled grenades and large-caliber sniper rifles. Also, with their doors and vents closed, these shelters are secure from a swarming SUAS attack, a problem that could present itself in future conflicts.[84]

More-robust designs are possible, and some European allies, the Saudis, Iraqis, and others have pursued such designs. Whether they make sense for an expeditionary air force is unclear,

---

[80] For more on the Ewa shelters, see Vick 2015a, pp. 46–47.

[81] Vick 2015a, p. 48.

[82] Fox, 1979, p. 71; United States Air Force, *Defense of Da Nang: Project CHECO Southeast Asia Report*, Hickam AFB, Hawaii: Headquarters, Pacific Air Forces, August 31, 1969, pp. 17–23.

[83] Vick 2015a, p. 48.

[84] Thanks to Colonel Michael Pietrucha, USAF, for sharing his experience with hardened aircraft shelters.

but they remain an option.[85] Alternatively, lighter ballistic protective materials (e.g., Kevlar) may offer options for deployable shelters that could provide protection against some classes of threats.[86] Rubb Military Buildings, for example, offers Kevlar panels that can be inserted into its expeditionary aircraft hangars.[87] Protective material science continues to make significant advances in the lightweight ballistic protection necessary for deployable shelters to be viable.[88]

On-Base Dispersal

On-base dispersal of assets has two primary purposes. First, it seeks to reduce the number of assets that can be damaged or destroyed by any one weapon. For example, if the main threat weapon has an effective radius of 100 m, on-base dispersal of aircraft would require ensuring that aircraft are never within 100 m of one another, thus forcing the attacker to use at least one weapon for every aircraft damaged or destroyed. One analysis found that, if fighter aircraft were separated by 440 m, an attacker would have to use more than five times as many missiles than for a base case with no dispersal.[89] The second purpose is to complicate the attacker's targeting problem by reducing the visual signature of the protected assets. Clusters of buildings, vehicles, and aircraft; a large fuel farm; and other assets are easier to detect than individual assets scattered around a base. On-base dispersal is often used in combination with CCD efforts, which we discuss in the next subsection. For example, dispersed aircraft have historically been concealed in tree lines or camouflaged with netting; individual vehicles can be parked next to buildings, making the vehicles harder to detect and attack.

On-base dispersal of aircraft became a common practice in World War II, and all combatants used it for the purposes detailed earlier.[90] It received renewed addition in the 1950s, when NATO and U.S. forces sought to minimize the air base vulnerability to nuclear attack.[91] Although generally less central to air base defense than in past years, on-base dispersal remains an important component and can be especially effective when combined with CCD efforts.[92]

---

[85] For more on such hardened aircraft shelter designs, see Vick, 2015a, pp. 50–51.

[86] See Stillion and Orletsky, 1999, pp. 32–35 for an assessment of the feasibility of such shelters.

[87] Rubb Military Buildings, "Aircraft Hangars," webpage, 2016.

[88] See National Materials Advisory Board, *Opportunities in Protective Material Science and Technology for Future Army Applications*, Washington, D.C.: National Academies Press, 2011.

[89] Fewer than ten missiles were required for the base case and more than 50 for the 440-m separation case. See Stillion and Orletsky, 1999, pp. 35–38.

[90] Kreis, 1988, p. 126.

[91] Lawrence R. Benson, *USAF Aircraft Basing in Europe, North Africa, and the Middle East 1945–1980*, Ramstein Air Base, Germany: Headquarters, U.S. Air Forces in Europe, 1981, p. 24.

[92] For current Air Force practices regarding on-base dispersal of assets, see AFTTP 3-32.34V3, 2016, pp. 51–52, and AFH 10-222, Vol. 1, *Civil Engineer Bare Base Development*, Washington, D.C.: Department of the Air Force, January 23, 2012, pp. 46, 49–50.

Camouflage, Concealment, and Deception

CCD measures are intended to hide friendly assets and activities from enemy observation and to deceive the enemy about the locations, capabilities, and intentions of friendly forces. CCD is applicable to land, air, sea, and space operations, both offensively and defensively. For air base defense, CCD complements on-base dispersal and distributed operations.

Combatant air forces have routinely used CCD to hide aircraft and even airstrips from the enemy and to deceive the enemy by using decoy airfields and dummy aircraft (often repurposing aircraft too damaged to fly). Perhaps the most successful deception in air base defense was the RAF's use of decoy airfields during World War II, attracting 440 Luftwaffe attacks, compared with the 430 directed against actual airfields in the UK.[93]

Decoy airfields were also used for offensive purposes. Prior to the Allied invasion of Lae, New Guinea, the USAAF needed to build an airstrip within range of Japanese bases at Wewak. To draw attention away from the airstrip they were covertly building at Tsili Tsili, "Allied engineers were visibly active at another inland site, Bena Bena. . . . This drew Japanese air attacks and a furtive land attack. With Japanese attention thus drawn to Bena Bena, the strip at Tsili Tsili accepted its first fighter unit and deployed a powerful force."[94] This strip provided critical fighter support for a major air offensive against the Japanese air bases at Wewak.

As noted in the previous subsection, dispersed aircraft were often concealed in tree lines or covered with camouflage nets during World War II, combining dispersal, camouflage, and concealment. One complementary deception technique that most combatants used was to place damaged aircraft or decoys along tree lines or in revetments to draw attacks away from operational aircraft.[95] Decoy aircraft have been used in other conflicts, including the Korean, Vietnam, and 1971 India-Pakistan wars.[96] During the Cold War, the Soviet Union used inflatable dummy aircraft as part of peacetime deception efforts.[97]

Looking to the future, the proliferation of increasingly capable remote-sensing devices on satellites, aircraft, and drones presents challenges for CCD, which now must fool multispectral sensors rather than just the human eye. Sophisticated decoys are available today that mimic the

---

[93] Seymour Reit, *Masquerade: The Amazing Camouflage Deceptions of World War II*, New York: Hawthorn Books, Inc., 1978, pp. 49–61. For a contemporary analysis of deception and concealment, see Jonathan F. Solomon, "Maritime Deception and Concealment: Concepts for Defeating Wide-Area Oceanic Surveillance-Reconnaissance-Strike Networks," *Naval War College Review*, Vol. 66, No. 4, Autumn 2013. Although focused on land force operations, Army doctrine offers useful insights. See Army Tactics, Techniques, and Procedures 3-34.39, *Camouflage, Concealment and Decoys*, Washington, D.C.: Headquarters, Department of the Army, November 2010.

[94] Bergerud, 2001, p. 629.

[95] Reit, 1978, p. 189.

[96] Vick, 2015a, p. 41.

[97] National Photographic Interpretation Center, "Inflatable Dummy Aircraft: Arkhanelsk/Talagi Airfield, USSR," Washington, D.C.: Central Intelligence Agency, June 16, 1983.

visual, IR, and radar signatures of aircraft, but these will increasingly have to be part of an integrated multidomain CCD strategy.[98]

Distributed Operations

Distributed air operations, sometimes referred to as *dispersed operations*, seek to enhance air operations by proliferating operating locations.[99] Increasing the number of operating locations has two benefits. First, increasing the number of airstrips makes it more difficult for the enemy to prevent friendly air operations through attacks on runways and other facilities. Second, spreading a given force across multiple locations reduces the aircraft density per location. For example, instead of a wing of fighters (roughly 72 aircraft) operating from one location, each of the wing's three squadrons (in this example, 24 aircraft each) would operate from its own airstrip. Thus, even a successful attack against one of these three airfields could damage or destroy 24 aircraft at most, as opposed to all 72 aircraft.[100]

Distributed air operations are not new. During World War II, combatant air forces routinely used dispersal or auxiliary fields to reduce their vulnerability. For example, Henderson Field on Guadalcanal was supported by two auxiliary fighter airstrips. Distributed operations became of great interest to SAC once the Soviet Union acquired nuclear weapons. Fearing that a surprise attack could wipe out the U.S. strategic bomber force, the Air Force commissioned many RAND studies assessing the cost-effectiveness of distributed bomber operations.[101] During the Cold War, the Soviet Union and Warsaw Pact built a vast network of bases, including 218 primary and 536 secondary airfields.[102] Similarly, by 1980, USAFE had access to 23 MOBs, five standby bases, and 72 colocated operating bases.[103]

An interesting twist to distributed operations involves not just spreading forces out but moving them among bases, so that some bases are occupied and others empty at various times. This increases enemy uncertainty about the location of friendly forces. This stratagem was first employed by the Germans in World War I.[104]

---

[98] INFLATECH offers a variety of military decoys. See INFLATECH, "Inflatable Military Decoys," undated.

[99] For an exploration of the challenges associated with distributed operations in the Pacific theater, see Miranda Priebe, Alan J. Vick, Jacob L. Heim, and Meagan L. Smith, *Distributed Operations in a Contested Environment: Implications for USAF Force Presentation*, Santa Monica, Calif.: RAND Corporation, RR-2959-AF, 2019. Also see William E. Pinter, *Concentrating on Dispersed Operations: Answering the Emerging Antiaccess Challenge in the Pacific Rim*, thesis, Air University, School of Advanced Air and Space Studies, Maxwell AFB, Ala.: Air University Press, 2007.

[100] Most likely, a significant fraction of the aircraft would be flying missions at any given moment, reducing the actual number of aircraft vulnerable to attack on the ground.

[101] The first of these was RAND Cost Analysis Section, *The Cost of Decreasing Vulnerability of Air Bases by Dispersal: Dispersing a B-36 Wing*, Santa Monica, Calif.: RAND Corporation, R-235, 1952.

[102] Benson, 1981, pp. 103–104.

[103] Vick, 2015a, p. 55.

[104] Kreis, 1988, p. 13.

Assessing the Versatility of Defensive Options Across Threats

As the Air Force considers its options to improve air base defenses, it will likely seek to minimize the number of separate capabilities and systems it must procure and sustain. Ideally, the Air Force could deploy a single multipurpose defensive system with significant capabilities across a wide range of threats. Outside passive defenses, there does not appear to be such a system.

Table 3.2 offers a broad consideration of the relative versatility of the defensive technologies and options discussed in this chapter. The assessment is subjective, based on a review of applicable military history and technology and policy assessments, as discussed earlier in the chapter. A green cell indicates significant capability to counter the specified threat; yellow reflects some current capability or promise for near-term future deployment; red indicates little to no capability against the threat. Note that green does not mean that the problem has been solved, only that the particular option has much to offer. More technical engagement- and campaign-level analyses are necessary to make such judgments—efforts that go well beyond the scope and remit of this report.

Table 3.2. Defensive Option Applicability Across Threats to Air Bases

	Swarming SUAS	Rockets, Mortars	Civil Aircraft	Combat Aircraft	Cruise Missiles	Ballistic Missiles	Hypersonic Weapons
RF jamming	Significant	Little to none	Some	Little to none	Little to none	Little to none	Little to none
High-powered microwave	Significant	Some	Some	Some	Some	Little to none	Little to none
Solid-state lasers	Some	Some	Some	Some	Some	Little to none	Little to none
Guns	Some	Significant	Significant	Some	Some	Little to none	Little to none
Short-range missiles	Little to none	Some	Significant	Significant	Significant	Little to none	Little to none
Long-range missiles	Little to none	Little to none	Significant	Significant	Some	Significant	Little to none
DCA	Little to none	Little to none	Significant	Significant	Significant	Little to none	Little to none
Passive defenses	Significant	Significant	Significant	Significant	Significant	Significant	Significant

NOTES: Colors indicate the degree of capability against the threat. Red indicates little to none; yellow indicates some current capability or future potential against threat; and green indicates significant capability against threat.

Passive systems have much to offer. The Air Force generally has full authority to pursue them on its own. They tend to be relatively simple and inexpensive and can be fielded in months and years rather than decades. For example, the typical hardened aircraft shelter for fighters found at U.S. Air Force bases in Europe and Asia has 18 in. of reinforced concrete protection. With its doors and vents closed, it provides protection from swarming SUAS, rockets, mortars, direct-fire infantry weapons, most submunitions, and near misses from unitary warheads. That said, passive defenses cannot protect aircraft or personnel in the open or prevent damage to operating surfaces. Although they are a good starting point, passive defenses must be complemented by active systems for a robust defense.

As Table 3.2 illustrates, it appears that a combination of active defenses will be needed to offer good capability across the range of threats. RF jamming offers a useful near-term defense against individual or swarming SUAS but may be defeated by more-autonomous UASs. Prototype high-power microwave (HPM) systems (e.g., THOR) offer much greater lethality against SUAS and may also be effective against rockets, mortars, aircraft, and cruise missiles. Solid-state lasers also look promising against the same set of threats. Both HPM systems and lasers are coded yellow for most of these threats because of uncertainties about their operational effectiveness. Both technologies are rapidly advancing in capacity but cannot yet be considered technologically mature. Specialized AAA guns deployed with partner-nation forces should be exploited, and the U.S. Army C-RAM capability is worth using when mortar and rocket threats are of concern. That said, investments in new AAA guns do not appear warranted at this time. Short-range missiles, as envisioned in IFPC or deployed in NASAMS, are well worth pursuing for air base defense.

Long-range missiles (e.g., Patriot, THAAD, SM-3, SM-6) are a critical part of theater integrated AMD but have significant limitations against cruise missiles and SUAS. DCA patrols for cruise-missile defense can be synergistic with ground-based SHORAD systems but are generally less effective against missile salvos.

Conclusion

This chapter assessed the operational utility of various defensive technologies, systems, and concepts in defeating some of the most worrisome threats to airfields. Our assessment was limited to technological and operational considerations regardless of the service that has been assigned responsibility for a particular mission or technology. For example, the U.S. Army is responsible for training, organizing, and equipping forces for ground-based air defense of both its own forces and rear-area installations. Thus, although short-range missiles have great potential for air base defense against cruise missiles, the Air Force does not believe it has the authority to procure these systems. As we will discuss in the next two chapters, ground-based air defense of fixed facilities is a low priority for the Army; consequently, no Army air defense units have significant counter–cruise missile capabilities, and no Army air defense units are dedicated to air base defense. In contrast, the Air Force is responsible for planning, programming, and budgeting for passive airfield defense. Although the Air Force still must build Office of the Secretary of Defense (OSD) and congressional support for such programs, it has the authority to develop plans and advocate for them.

In sum, the relative performance of air base defense options is only one consideration; Air Force leaders must also consider what options the service can undertake unilaterally as opposed to those that require either new authorities or changes in the priorities of other organizations.

The next three chapters explore air base defense from the perspective of service R&F as we seek to understand why air base air defense was taken away from the Air Force; how the Army has understood its responsibilities; what issues have caused the most friction between the two services; and, in light of R&F constraints, what COAs the Air Force might pursue to improve air base defense capabilities.

4. Roles and Missions: Key West to the Vietnam War

The debates commonly associated with the roles and missions of the U.S. armed forces are long standing and have a storied history. Disputes among the services over who should command forces and direct operations were common in World War II. Wartime decisions on these questions were expedient and offered no fundamental framework or logic for the postwar world. The creation of an independent Air Force and DoD and the new reality of nuclear weapons further exacerbated the existing interservice tensions and rivalries.

Common usage of the term *roles and missions* within the military dates back at least to the National Security Act of 1947,[1] the landmark legislation that unified the armed forces under the National Military Establishment and, later, DoD.[2] Today the expression—and the associated debates—continue. The Senate version of the 2018 National Security Act included a provision calling for the Secretary of Defense to "submit to the congressional defense committees a report setting forth a reevaluation of the highest priority missions of DoD, and of the roles of the Armed Forces in the performance of such missions."[3]

Importantly, the broader question of which service should have ownership of which functions and responsibilities has huge implications for the defense of air bases—a function that spans multiple threat dimensions. In this chapter, we examine service roles and missions—with a distinct focus on air defense and the defense of air bases in particular. While a broader and lengthier treatment of the services debates related to roles and missions is beyond the scope of this report, we do offer a bit in the way of background.

The roles and missions debate is tracible back to the 1948 Key West Conference, where the first Secretary of Defense, James Forrestal, convened the service chiefs to clarify service responsibilities. The conference concluded with a paper, commonly referred to as the Key West Agreement, that established primary and collateral functions for each service.[4] As a compromise document between the services, the Key West Agreement was intentionally vague, allowing

---

[1] Pub. L. 253, National Security Act of 1947, July 26, 1947.

[2] In fact, the term predates the National Security Act of 1947, having appeared often in early post–World War II debates over the military's unification after the war and particular service responsibilities.

[3] Ryan Evans, "Call for Articles: The Military Roles and Missions Analysis That America Deserves," War on the Rocks website, August 15, 2018; Sec. 1041 of Pub. L. 115-232, 2018.

[4] James V. Forrestal, "Functions of the Armed Forces and the Joint Chiefs of Staff," appendix to "Note by the Secretaries to the Joint Chiefs of Staff," Washington, D.C.: U.S. Department of Defense, April 21, 1948. The document codifies that the Air Force is responsible for the air domain, Army for the land, and Navy for the sea. Beyond these basic distinctions, this document does not provide additional specifics to adjudicate many long-standing disagreements about roles and missions.

wide variations in interpretation in some areas. Moreover, it did not mention Air Force ground combat forces or assign the function of air base defense to the Air Force.[5]

Important to the evolution of air base defense, the Key West Agreement also did not address guided missiles. This gap allowed each service to develop SAMs independently—a factor that would spawn new controversies in the decades to follow. The Army had owned the ground-based antiaircraft mission since World War I, when the Army Coast Artillery branch was given its mission:

> The Coast Artillery was the primary source of manpower and expertise for the fledgling Antiaircraft Service by virtue of its experience with plotting firing solutions for moving two-dimensional seagoing targets near the coasts and harbors of the United States. Based on this experience, the Army felt that it would be a relatively simple transition to firing on targets in the three dimensions in the air.[6]

During World War II, Army AAA capabilities expanded greatly, but AAA remained part of coastal artillery. In 1947, the Army combined coastal, AAA, and field artillery into a single branch. As SAM technology advanced, the Army recognized that its antiaircraft capabilities could be greatly improved by incorporating missiles and that SAMs might make AAA obsolete.[7] Thus, by leveraging its control of the AAA mission, the Army argued for responsibility for the development of all SAMs. The Air Force would counter this position, arguing that its control of strategic missiles and pilotless aircraft implied its own responsibility in this area. These debates—over missiles and who had ultimate responsibility for defending air bases—would unfold with intensity beginning in the 1950s. In the next section, we trace this history and the debates, with a particular focus on the roles and missions aspect of defending U.S. Air Force bases.

Air Force–Army Debates over Air Base Defense: 1950s

The 1950s were marked by President Dwight Eisenhower's New Look military strategy during a period of growing concern about Soviet military advancements and technological improvements. The objective of the policy was to meet U.S. Cold War military obligations while avoiding excessive defense expenditures. The emphasis was therefore on strategic nuclear weapons, especially strategic airpower, within a slimmed-down military establishment. This redirection also prioritized the protection of the continental United States (CONUS) over other global military commitments. Credible deterrence was predicated on a strategy of retaliation with nuclear weapons if attacked. In the face of such fiscal constraints, interservice rivalries reached

---

[5] For insights related to Key West and the defense of air bases, see Erik K. Rundquist, "A Short History of Air Base Defense: From World War I to Iraq," in Caudill, 2014, p. 9.

[6] Hamilton, 2009, p. 25.

[7] See Hamilton, 2009, for a detailed history of U.S. Army air defense artillery.

new levels. Disputes about roles and missions moved to front and center and were even on full display in the halls of Congress as services sought to establish and defend appropriation requests to support their programmatic objectives.

The difficulties notwithstanding, the early part of the decade was punctuated by an agreement between the Army and Air Force over the protection of the CONUS. When the Soviet Union developed a nuclear capability in 1949, the U.S. military began prioritizing air defense at home. Until 1949, however, the Air Force was responsible for CONUS air defense. In theory, to defend CONUS from air attack, the Air Force and Army would need to integrate Army AAA battalions and Air Force interceptor squadrons.

Recognizing the growing need for better integration of Air Force and Army assets in CONUS air defense, Gen Hoyt Vandenberg, CSAF, and GEN J. Lawton Collins, Chief of Staff of the Army (CSA), forged an agreement that set out the respective responsibilities for the two services. The 1952 agreement covered six key aspects of the control of U.S. Army antiaircraft units by USAF air defense commanders:

1. Each echelon of USAF air defense command would include an Army AAA element, with its commander serving as the antiaircraft advisor to the USAF air defense commander.
2. The Air Force commander will promulgate rules of engagement (described in the agreement) for AAA units for use throughout the United States.
3. After declaration of a state of emergency by higher authorities, the readiness condition for each AAA unit will be established by the USAF air defense commander in conformance with a JCS approved plan.
4. In the absence of an approved JCS plan, the Department of Army and Department of Air Force will determine which geographic areas are to be provided AAA defense.
5. The Army and Air Force will mutually determine locations for AAA units but Army AAA officers will command those units and determine their specific tactical siting.
6. Operational control over both Army and Air Force air defense units will be exercised by the Commander of the Air Defense Division.[8]

A mutually agreed on joint arrangement to defend the homeland would not prevent difficulties from emerging during the 1950s. Part of the problem that the department and the services had to deal with stemmed from new technological developments of the era.

Advancements in jet engines, ballistic missiles, and nuclear weapons ushered in an era of vulnerability for the United States, no longer protected behind two vast oceans. To deal with these complexities, the U.S. military began to develop new airplanes, missiles, and air defense systems. Because these systems were considerably more complex than their predecessors, they invited new questions. How did these programs complement or one another? How could they be integrated? Which services should have responsibility for which capabilities? How could the

---

[8] Vandenberg-Collins Agreement, August 1, 1950, reproduced in Richard I. Wolf, *The United States Air Force: Basic Documents on Roles and Missions*, Washington, D.C.: Office of Air Force History, 1987, pp. 221–222.

U.S. military avoid waste and unnecessary duplication in an era of fiscal tightening? How should different systems and capabilities be prioritized?[9]

Within the context of these broader debates, disagreements between the U.S. Air Force and the U.S. Army would emerge, especially over the air defense function. A critical aspect of these differences centered around the development of new guided-missile technologies. In spite of its primary responsibility for missile research after World War II, the Air Force faced competition from other services, especially the Army, in the guided missile field. A 1949 review of the guided-missile effort resulted in the Air Force attaining singular responsibility for the development and employment of long-range strategic missiles, while the Navy and Army retained interests in short-range missiles. And in shared mission areas, such as air defense, all the services developed and employed guided-missile technology.[10] This overlapping arrangement did little to ameliorate turf wars between the services. In particular, although the Key West Agreement afforded the Air Force the primary role of air defense, the Navy and Army maintained collateral duties in this mission.[11]

Conflicting service priorities also complicated efforts to create a unified defense command. Early in 1954, the Chairman of the Joint Chiefs of Staff (CJCS), ADM Arthur Radford, obtained presidential approval to establish such a command. The Air Force had resisted prior efforts to create a unified command, only to reverse its position after assurances from Admiral Radford and the Secretary of Defense that the command would enhance the Air Force's role in air defense. The Army demurred on the establishment of the new command. It remained fearful that its primary role in air defense (artillery and SAMs) would be subordinated to Air Force needs.

The Joint Chiefs of Staff (JCS) dismissed the Army's protest. In summer 1954, the Continental Air Defense Command was officially established. The four-star command was to be headed by an Air Force general, and the Air Force became its executive agent, with Army and Navy forces serving as joint command components. By 1957, the integration of Canadian and U.S. air defenses brought about the establishment of the North American Air Defense Command, headed by a U.S. Air Force general.[12] The dispute between the Army and the Air Force over the establishment of the Continental Air Defense Command was merely one chapter of unabated interservice bickering during the decade.

---

[9] For discussion of these and related questions, see Joshua Klimas, *Balancing Consensus, Consent, and Competence: Richard Russell, The Senate Armed Services Committee & Oversight of America's Defense, 1955–1968*, dissertation, The Ohio State University, 2007, pp. 186–187.

[10] Warren A. Trest, *Air Force Roles and Missions: A History*, Washington, D.C.: Air Force History and Museums Program, 1998, pp. 160–161. For an Army perspective, see also Mark L. Morgan and Mark A. Berhow, *Rings of Supersonic Steel: Air Defenses of the United States Army 1950–1979*, rev. 3rd ed., Bodega Bay, Calif.: Hole in the Head Press, 2010.

[11] The Army retained its AAA responsibility, while the Navy maintained sea-based air defense of coastlines. See Trest, 1998, p. 161.

[12] Trest, 1998, p. 162.

One of the most heated debates between the two services centered on air defenses. In the mid-1950s, the Army was pursuing the development and modernization of its Nike antiaircraft missile system. The Nike Ajax had been deployed since 1953, but the Army soon began work on Nike-B, later renamed the Nike Hercules. The Nike Hercules was faster and more accurate and had a longer range than its predecessor. In 1956, the Air Force sought authorization to construct its own land-based missile system to provide point defense for its SAC bases. The Air Force wanted a longer-range weapon than the Nike Ajax and saw the U.S. Navy Talos missile as ideal for its needs. The two services pushing for different air defense systems set the stage for a rather dramatic series of testimonies before the Senate Armed Services Committee in summer 1956 concerning appropriations for fiscal year (FY) 1957.

Early in 1956, under the leadership of Senator Stuart Symington, a former Secretary of the Air Force, the Senate Armed Services Committee formed a special subcommittee to examine the challenges U.S. airpower faced. Beginning in the spring, the committee held a special hearing on airpower that would last into the summer. The senate hearings revealed divergent opinions on numerous issues related to air defense, including roles and missions. The purpose of the committee was to study U.S. strategic nuclear delivery systems (bombers and missiles) and air defense capabilities in the context of improved Soviet bomber capabilities.[13] Inevitably, the Nike-Talos controversy would emerge during the months of testimony as the senators sought out specifics regarding both systems.[14]

In fact, the Chief of Naval Operations, ADM Arleigh Burke, broached the subject during his own testimony before the committee, even though the nominal topic of his hearing was unrelated to the dispute. Describing the most difficult problems the military confronted, the admiral would say that "the major problems at the moment" dealt with roles and missions, specifically missiles: "Who should have certain kinds of missiles, the Army or the Air Force? That is not an easy problem to solve."[15] Senator Symington's response to the admiral's comments revealed a growing frustration with interservice rivalry:

> You have brought up the current missile problem between the Air Force and the Army. It is a big problem. In my opinion neither service is suffering the most. The fellow who is suffering the most is the taxpayer. The services get into an

---

[13] Klimas, 2007, p. 190.

[14] The previous year, the Army had sought $160 million from Congress to improve air defense construction and activities worldwide. A Senate Armed Services Committee subcommittee member, however, remained skeptical that the Nike Ajax possessed sufficient range to intercept all Soviet bombers. Moreover, he was concerned that by the time the Army fully fielded the Ajax, it would be obsolete, and the Army would find itself petitioning Congress for appropriations to field a new system. See Klimas, 2007, p. 195.

[15] U.S. Senate, "Study of Airpower," hearings before the Subcommittee of the Air Force of the Committee on Armed Services, Part XVIII, 84th Cong., 2nd Sess., June 18 and 27, 1956b, p. 1381.

argument on Nike and Talos, and now we hear that the services say both are needed.[16]

The senator was correct in his assertion that the respective parties in the dispute maintained that both systems were necessary. In an effort to resolve the disagreement and justify their positions, the Air Force and Army claimed that the Nike Ajax and Talos fulfilled two separate missions. The Army specifically pointed to the function of point defense, for which the Ajax was a highly capable system. For its part, the Air Force emphasized its primary role in the broader mission of area defense.[17] The extended range of the Talos over the Ajax made it the preferable weapon. By separating point from area defense, the two services maintained that there was no duplication in the respective weapon systems.[18] GEN Maxwell Taylor, CSA, would summarize these positions in his own testimony when pressed about the service chiefs' positions on the two missiles:

"The general concept was the Army was interested in extending its traditional antiaircraft artillery role, which is largely point defense of vital targets, whereas the Air Force's legitimate interest was more in the interceptor role, so that the missiles they would go for would perform interceptor-type missions. It was felt that was a commonsense limit at the time, although we specifically agreed that this could be changed, and there was no desire on either services' part, really, to restrict the capability of weapons which is to the interest of the United States to develop."[19]

The senators, however, saw matters differently. They remained unpersuaded by the services' arguments, offering the following conclusion on the matter:

> The committee believes that the proponents for each weapon system are dedicated and sincerely patriotic individuals, whose sole interest is to provide the best possible national defense. The committee concluded that both the Army and the Air Force are assigned overlapping roles and missions in the antiaircraft and continental air defense fields. While the Air Force views its mission as one of area defense, and the Army views its as one of perimeter or point defense, it is clear that definite and urgent need exists for the Department of Defense to quickly and positively clarify the specific responsibility of each service. The committee believes that unless concise responsibilities are assigned, duplication of weapon systems costing in the multi-billion-dollar range might result, and that such duplication would obviously be too costly as well as inexcusable from a military standpoint.[20]

---

[16] U.S. Senate, 1956b, p. 1381.

[17] The Air Force's emphasis or concern may well have begun prior to the missile debate highlighted here. However, the concern certainly helped the Air Force justify its position regarding the Talos missile system.

[18] Klimas, 2007, p. 199.

[19] U.S. Senate, "Study of Airpower," hearings before the Subcommittee of the Air Force of the Committee on Armed Services, Part XVII, 84th Cong., 2nd Sess., June 18 and 25, 1956a, p. 1284.

[20] U.S. Senate, *Authorizing Construction for Military Departments*, Senate Committee on Armed Services report, 84th Cong., 2nd Sess., July 25, 1956c, p. 4.

The committee also rebuked the U.S. military for its inability to resolve its own interservice issues: "Congress should not be placed in the position of defining roles and mission even by inference unless such is accomplished by specifically designed legislation, supported by concrete recommendations on the part of responsible Department of Defense officials."[21]

The Nike-Talos episode eventually receded into history at the end of 1956 with the Senate Armed Services Committee's decision to shift Talos development to the Army. The committee specifically called for further testing of the two systems to determine their relative merits. But Secretary of Defense Charles Wilson did not honor such demands. Unsurprisingly, the Army effectively killed the Talos initiative, deciding to proceed with the Nike-Hercules program instead. In retrospect, this was likely the right decision. The feasibility of the Talos missile for a land-based role was unclear at the time, and the Hercules would go on to serve as the military's primary SAM until it was replaced by the Patriot in the 1980s.

In November 1956, Secretary Wilson issued a formal memorandum to the members of the Armed Forces Policy Council to clarify roles and missions to improve the operation of DoD.[22] The council was created to advise the Secretary of Defense on policy pertaining to the armed forces. The Secretary of Defense could also use this body to investigate and report on issues they needed.[23] Acknowledging the inherent difficulty of crafting changes in roles and missions, the secretary noted that the roles and missions laid out at Key West did not require basic alteration but that the advent of new weapons and strategic concepts had "pointed up the need for some clarification and clearer interpretation of the roles and missions of the armed services."[24] Specifically on air defense, Wilson noted that consideration was given to distinguishing between Air Force and Army responsibility for surface-to-air guided-missile systems for the defense of CONUS on the basis of both area and point defense. To the Army, the secretary assigned "responsibility for the development, procurement and manning of land-based surface-to-air missiles systems for point defense." He further specified that the current systems in this category were Nike I, Nike B, and the land-based Talos.[25] The Air Force—having officially lost its bid for the Talos—was assigned parallel responsibilities for land-based systems for area defense. The Air Force's Bomarc, a 400-nm-range missile designed to supplement fighters in the area defense role, was the current system in this category.[26] Importantly, the Wilson memo assigned "sole

---

[21] U.S. Senate, 1956c, p. 5.

[22] Charles E. Wilson, "Clarification of Roles and Missions to Improve the Effectiveness of Operation of the Department of Defense," memorandum for members of the Armed Forces Policy Council, Washington, D.C.: November 26, 1956, as reproduced in Wolf, 1987.

[23] The membership and purpose of the council are outlined in U.S. Code, Title 10, Armed Forces, Subtitle A, General Military Law, Pt. I, Organization and General Military Powers, Ch.7, Boards, Councils, and Committees, Sec. 171, Armed Forces Policy Council.

[24] Wilson, 1956.

[25] Wilson, 1956.

[26] For more technical details on the Nike, Talos, and Bomarc systems, see Berhow, 2005.

responsibility" for the operational employment of land-based IRBM systems to the Air Force rather than to the Army.[27] The memorandum also confined Army use of SAMs to the 100-mi range limit imposed forward of the front lines—a limitation Army planners believed far too restrictive for future battle zones.[28] Eventually, the range limit disappeared, giving the Army control of all land-based SAMs, whatever their range.

1960s and 1970s: Vietnam—Defending Bases During War

This section briefly details the ground threat to Air Force bases in Vietnam during the 1960s and early 1970s. The war in Vietnam represented the first time that the U.S. Air Force operated from bases within territory subject to enduring insurgent threats. These included mortars, artillery, rockets, machine gun and small arms fire, and sapper raids. There were more ground attacks on air bases in Vietnam than in any previous U.S. conflict. From 1964 to 1973, the Viet Cong and the NVA attacked U.S. Air Force MOBs 475 times, destroying 99 U.S. and Vietnamese aircraft and damaging a further 1,170. Additional attacks on Air Force, Marine Corps, and Republic of Vietnam Air Force facilities in Vietnam and Thailand brought the total aircraft destroyed to 375.[29] While the destruction of aircraft on the ground was only about 11 percent of total losses, the U.S. Air Force lost more fixed-wing aircraft to enemy ground attacks than it did in air-to-air combat during the conflict.[30] Moreover, the sheer volume of attacks on U.S. and Vietnamese MOBs—more than 120 in 1968 alone—added to a heightened sense of vulnerability.

U.S. facilities proved so vulnerable to attacks partly because of a failure to take the determination and resourcefulness of the adversary seriously and partly because of a decision to rely on South Vietnamese forces for rear-area security, even though their deficiencies in training, leadership, and equipment were well known.[31] The U.S. unified command had, as early as 1961,

---

[27] Wilson, 1956.

[28] See Trest, 1998, p. 174, which also notes that "Wilson's memorandum allowed the Army to continue limited feasibility studies on IRBMs, but insisted that the missiles be turned over to the Air Force. Jupitor [sic] and Thor missiles were assigned to SAC when they became operational in 1958."

[29] Vick, 1995, p. 68.

[30] In total, all services lost 3,322 fixed-wing aircraft during the war, according to Chris Hobson, *Vietnam Air Losses: United States Air Force, Navy and Marine Corps Fixed-Wing Aircraft Losses in Southeast Asia, 1961–1973*, Hinckley, UK: Midland Publishing, 2001. See also Vick, 1995, p. 69.

[31] Although MACV's threat assessments failed in general, there were notable exceptions. Maj Gen Joseph H. Moore, the 2nd Air Division Commander, convinced Pacific Air Forces (PACAF) Headquarters to return one of the two B-57 squadrons at Bien Hoa to its home base at Clark AB, Philippines because of the mortar threat. This happened just nine days before the Viet Cong attack. Moore cited both PACAF and 2nd Air Division studies that concluded that a Viet Cong mortar attack could destroy 50 percent of the aircraft on the ramp. See CHECO Office, *Historical Background to Viet Cong Mortar Attack on Bien Hoa: 1 November 1964*, Honolulu, Hawaii: Headquarters 2nd Air Division, November 9, 1964, pp. 7–9; Richard R. Lee, *7AF Local Base Defense Operations, July 1965–December 1968*, Hickam AFB, Hawaii: Headquarters Pacific Air Forces, July 1, 1969, p. 27; and

taken the position that the defense of U.S. resources in Vietnam was the responsibility of the Vietnamese. This policy was reaffirmed by U.S. Military Assistance Command, Vietnam (MACV) Headquarters.[32] This position was directly at odds with the fact that U.S. intervention in the conflict was largely precipitated by the inability of the government of Vietnam to defend itself. As the Army of the Republic of Vietnam (ARVN) progressively weakened and as the Viet Cong grew stronger, the folly of this policy would soon become self-evident. In November 1964, Viet Cong guerrillas struck Bien Hoa AB, subjecting it to a 30-minute barrage of 81-mm mortar rounds in which five B-57 bombers were destroyed; another eight were heavily damaged; and seven were lightly damaged, in addition to other aircraft losses. In short, the entire B-57 squadron was taken out of action. Four U.S. and two South Vietnamese personnel also died in the standoff attack, and 72 were wounded.[33]

The affair demonstrated beyond doubt that ARVN defense measures were inadequate and uncoordinated.[34] Moreover, policy on defending U.S. facilities and bases in Vietnam was insufficient in the face of the growing ground threat. By April 1966, every U.S. Air Force MOB in Vietnam had been attacked.[35] After the 1964 attack on Bien Hoa, the MACV commanding officer, GEN William Westmoreland, called for improvement in organization, integration, and alert posture of reaction forces (infantry, artillery, and air), and for stepping up passive measures, such as greater dispersal of aircraft and more shelters. Although MACV leadership continued to press its Republic of Vietnam counterparts to shore up efforts to defend fixed facilities in the face of expanded enemy offensive operations, Air Force commanders considered Republic of Vietnam proficiencies for the task to be lacking.

Gen Hunter Harris, Commander-in-Chief, PACAF, was particularly alarmed. In November 1964, the same month as the Bien Hoa attack, he suggested that the United States use its Marines or Army forces to secure and control an 8,000-m² area around Da Nang, Bien Hoa, and Tan Son Nhut. General Westmoreland rejected this proposal outright because the President and senior military leadership still saw the combination of a coercive air campaign with advising and equipping of South Vietnamese forces as the preferred strategy. Deploying U.S. ground combat forces was simply not a serious option in fall 1964. Westmoreland did concede to a request for the deployment of 300 more security police to South Vietnam for internal security.[36]

Frederick Torgerson, *Parked Aircraft Vulnerability to Mortar Attack*, Hickam AFB, Hawaii: Headquarters Pacific Air Forces, September 9, 1964.

[32] Roger P. Fox, "Air Base Defense: An Appraisal," *Aerospace Commentary*, Vol. V, No. 1, Winter 1972, p. 19.

[33] Vick, 2015a, pp. 25–26. For a more-detailed after-action report on the Bien Hoa attack, see Pacific Air Forces, *Follow-Up to Bien Hoa Mortar Attack*, Project CHECO staff report, Hickam AFB, Hawaii: Headquarters Pacific Air Forces, December 1965a.

[34] Fox, 1979, p. 16.

[35] Vick, 1995, pp. 76–77.

[36] Fox, 1979, p. 17.

In just a few months, however, Westmoreland's assessment of the threat to at least one air base—the Marine Corps–Air Force airfield at Da Nang—had fundamentally changed. According to Westmoreland biographer Samuel Zaffiri:

> His most immediate problem at the start of 1965 was the airfield at Da Nang, from which many of the Rolling Thunder missions came. Fearful that the Vietcong might do the same thing to it that they had done to the strip at Bien Hoa, on February 22, he ordered his deputy, Major General Throckmorton, to fly up to Da Nang and inspect its security. Throckmorton returned the next day with grim news: The security was lax and the base in imminent danger of being overrun by the twelve VC [Viet Cong] battalions in the immediate area. Throckmorton recommended that a Marine expeditionary brigade of three battalions be landed there as soon as possible. Westmoreland agreed with Throckmorton's analysis but, preferring to keep American ground forces at a minimum, cabled Washington and told them he wanted to put two battalions in Da Nang and hold two more in reserve in ships just offshore.[37]

In his autobiography, Westmoreland discusses the concern that he and, especially, Ambassador Maxwell Taylor had regarding the introduction of U.S. combat forces:

> I saw my call for marines at Da Nang not as a first step in a growing American commitment but as what I said at the time it was: a way to secure a vital airfield and the air units using it, for which I saw no alternative, an airfield essential to pursing the adopted strategy [of a coercive air campaign against North Vietnam]. . . . Admiral Sharp at CINCPAC [Commander-in-Chief, U.S. Pacific Command] agreed with my two-battalion proposal, deeming it "an act of prudence which we should take before and not after another tragedy occurs." Washington on February 26 approved it, subject to South Vietnamese concurrence.[38]

The Marines landed on March 8, 1965. Westmoreland's reaction is instructive, suggesting that he had a deeper appreciation of the rear-area security problem than he is given credit for:

> Although my concern about the Da Nang air base was alleviated, I remained disturbed about possible enemy action against other bases, notably a U.S. Army communication facility and a small airfield at Phu Bai, near Hue, not a good field but at the same time the best we had north of the Hai Van Pass, and bases at Bien Hoa and Vung Tau. I asked Washington for an Army brigade for Bien Hoa and Vung Tau and another battalion of the III Marine Amphibious Force to come ashore and move to Phu Bai.[39]

---

[37] Samuel Zaffiri, *Westmoreland: A Biography of General William C. Westmoreland*, New York: William Morrow and Company, Inc., 1994, pp. 131–132.

[38] William C. Westmoreland, *A Soldier Reports*, Garden City, New York: Doubleday & Company, 1976, p. 123. For more on the debate over the introduction of U.S. forces, see Zaffiri, 1994, pp. 131–145, and Westmoreland, 1976, pp. 119–135.

[39] Westmoreland, 1976, p. 125.

By May, the Viet Cong had launched its summer offensive with a series of heavy attacks in three provinces, more evidence of its growing strength. After the Viet Cong had routed an ARVN regiment at Ba Gia,

> Westmoreland cabled CINCPAC and the JCS and told them pointedly that the enemy offensive was so fierce that ARVN battalions were being destroyed faster than he could replace them . . . . Westmoreland concluded the cable with the observation that he could only save South Vietnam from total collapse if he were given more than double the troops already in the pipeline, for a total of 180,000 men, or forty-four battalions.[40]

Westmoreland got his 44 battalions. Although the Air Force initially viewed this turn of events with optimism, this enthusiasm faded as Viet Cong and VNA attacks on air bases increased in frequency and intensity over the coming months and as the forces promised for rear area security were shifted to offensive operations. Some airmen saw this as a bait and switch, arguing that Westmoreland had justified 21 of the 44 battalions for rear area security but now was reneging on that commitment.[41] Westmoreland's view was, not surprisingly, different, arguing that, from the beginning, he had envisioned a multiphase approach to operations:

> On May 8 I forwarded to Washington my concept of how operations were to develop. In Stage One the units were to secure enclaves, which I preferred to call base areas, and in defending them could operate out to the range of light artillery. In Stage Two the units were to engage in offensive operations and deep patrolling in co-operation with the ARVN. In Stage Three they were to provide a reserve when ARVN units needed help and also conduct long-range offensive operations.[42]

Although the course of the war did not follow the happy trajectory in which ARVN forces would become the prime offensive forces, historian Dale Andrade has argued that Westmoreland was right about the need for an early shift to the offensive:

> Westmoreland's strategy worked in the sense that it saved South Vietnam from immediate defeat, pushed the enemy main forces away from the populated areas, and temporarily took the initiative away from the Communists. . . .

> These operations badly hurt the Communists. According to one analysis, "American search-and destroy missions disrupted the planned operations of the Viet Cong and thus made it more difficult for the Communist to seize the initiative. This became increasingly obvious to Hanoi in late 1965 and early 1966."[43]

---

[40] Zaffiri, 1994, pp. 140–141.

[41] Fox, 1979, pp. 21–22, for example, makes that argument. See pp. 17–27 for a more-general critique of Westmoreland's handling of air base security.

[42] Westmoreland, 1976, p. 135.

[43] Dale Andrade, "Westmoreland Was Right: Learning the Wrong Lessons from the Vietnam War," *Small Wars and Insurgencies*, Vol. 19, No. 2, June 2008, p. 161. The quote within the quote is from Patrick J. McGarvey, ed.,

These arguments, however sound at the national or MACV level, offered little comfort to Air Force leaders who had become increasingly frustrated by failure of the U.S. Army and Republic of Vietnam to provide adequate forces for air base defense. General Harris presented a pessimistic assessment of base defense organization in a personal letter to Gen John P. McConnell, CSAF, describing it as inadequate, lacking clearly identified responsibilities, and not under centralized control.[44] He asked that McConnell take the matter up with the JCS. In his response, McConnell noted that he shared Harris' "views and concerns regarding the deficiencies which appear to exist in the external base security arrangement in RVN [the Republic of Vietnam]." He added: "It is my intention to do everything possible to hold the Army to its mission of providing adequate external base security." He also indicated that, if an Inspector General inquiry merited such action, he would "address the problem to the JCS or the Army."[45]

General McConnell would indeed take the issue up with the JCS. In September, he proposed that the JCS send a message requesting that CINCPAC reexamine the base defense problem to ensure that U.S. base protection would be accorded first priority, that U.S. ground forces would defend base perimeters and offer protection from Viet Cong infiltration, and that external area defense operations would be sufficient to eliminate the possibility of mass attack and minimize enemy capability to conduct standoff attacks.[46] The JCS did not accept the proposal. This was the last time the Air Force would refer the issue to the JCS.

By the end of 1965, General Westmorland had clarified his position on the issue of base defense in a letter to the Commander of the 2nd Air Division:

> In order to provide a high level of security to airfields, it would be necessary to deploy a large number of U.S. infantry elements in a defensive role. Obviously, this cannot be done and, at the same time, go over to the offensive and destroy the VC. Therefore, I desire that all Service units and all forces of whatever Service who find themselves operating without infantry protection will be organized, trained and exercised to perform the defense and security functions which I have discussed.[47]

Westmoreland felt that he already was providing more forces for rear-area security than he could afford, with 50 percent of U.S. ground forces "tied down in securing base areas."[48] As the

---

Visions of Victory: Selected Vietnamese Communist Military Writings, 1964–1968, Stanford, Calif.: Hoover Institution on War, Revolution, and Peace, 1969, p. 5.

[44] Fox, 1979, p. 26.

[45] CSAF, AFCCS 76672, 0113502, message to CINCPACAF, September 1965. An Inspector General report in response to the CSAF's request would confirm an Army interest, emphasis, and cultural preference for offensive operations. Pacific Air Forces, "The Role of Aerospace Security Forces in Limited War Operations," memorandum to Headquarters, U.S. Air Force, December 1, 1965b.

[46] Fox, 1972, p. 23.

[47] William Rector, *The Role and Mission of Air Base Defense in a Counterinsurgency War*, Maxwell AFB, Ala.: Project Corona Harvest, Aerospace Studies Institute Air University, May 1970, pp. 25–26.

[48] Quoted in Fox, 1979, p. 22.

fighting escalated and as North Vietnam began to send much-more-capable mainline NVA battalions south, MACV struggled to field all the offensive forces that Westmoreland deemed necessary for victory. He simply could not afford to waste high-quality maneuver forces on less-demanding rear-area security duties.

Westmoreland's position was that the services should develop in-house units to provide needed defenses. This applied to all services, not just the Air Force. For example, the following year, MACV denied a request from the commanding general of U.S. Army forces in Vietnam that the 1st Infantry Division be given the mission of defending and/or participating in the defense of the ammunition supply depot in Lang Binh. In its rejection, headquarters reiterated to the Army general that its policy on base defense was, in essence, "that defensive and security functions must be performed by installation commanders with forces available."[49] Suffering manpower shortages, the MACV policy was an acceptance of a calculated risk.

While the policy applied equally to all the services, it would prove most challenging for the Air Force. Although only a small percentage of Army personnel in the 1960s were fully trained infantry, all soldiers received a level of ground combat training well beyond that given to airmen at the time. Beyond that, armies are built around movement, and movement entails some degree of self-sufficiency, including providing for one's own security in the field, whether the unit is infantry, communications, logistics, or transportation. This was even more true in the Marine Corps, where every Marine is considered an infantryman first, whatever his or her actual specialty. Both ground forces possessed a variety of organic weapons helpful to installation defense. For example, Army transportation companies had .50 caliber machine guns (for some trucks) as part of their normal Table of Organization and Equipment. Thus, even if an Army rear-area facility had no infantry companies assigned to its defense, it possessed an intrinsic capability for defense that was superior to that of the Air Force. Additionally, both the Army and the Marine Corps routinely purchased infantry weapons, ammunition, and related equipment in vast quantities, which eased the provision of materiel to support forces that had to provide for their own security.

The closest the Air Force had to combat infantry were its security police squadrons, which were neither trained nor equipped for the mission. The Air Force, along with the other services, was forced to create base defenses without much help from MACV.[50] Consequently, the Air

---

[49] U.S. Military Assistance Command Headquarters, Vietnam, "Request for Security Forces," message to Commanding General, U.S. Army, Vietnam, January 12, 1966.

[50] To be fair, U.S. Army, U.S. Marine Corps, Vietnamese Air Force security forces, and ARVN forces all made contributions to air base defense. For example, Marine Corps ground forces provided the bulk of the defenses of the joint Marine Corps–Air Force base at DaNang. Also, U.S. Army liaison officers and trainers worked closely with ARVN and regional Vietnamese ground forces to assist in air base defense. The quality of Vietnamese forces varied greatly, however, and they were generally not counted on for the bulk of defenses. See Fox, 1979, pp. 115–124; Rebecca Grant, "Safeside in the Desert," *Air Force Magazine*, May 6, 2008, p. 47.

Force initiated efforts to train and equip security police for base defense.[51] By the end of 1965, the Air Force had hastily dispatched 2,100 security police to South Vietnam.[52] Between 1965 and 1968, the Air Force made a series of ad hoc adaptations in the training, organization, and equipping of its Security Force squadrons to meet this new challenge. Following battalion-size attacks against U.S. Air Force bases during the 1968 Tet Offensive, the Air Force sought to enhance its capabilities through the creation of its own light infantry battalion under what became known as Operation Safe Side.[53] The number of permanently assigned forces peaked at around 4,700 security police in 1969.[54]

Manpower and personnel gaps proved to be one of the greatest early impediments to base defense. To deal with the shortages, the Air Force often haphazardly made officer assignments. One officer's after-action report illustrated the point:

> I came to Vietnam as a security police officer with no idea of what a security
> police officer was supposed to do. I was taken from another career field, given no
> training and shipped to one of the most important bases in Southeast Asia where I
> was to be responsible for the protection of over 5,000 lives and millions of
> dollars in vital equipment. Even though the base and I have survived so far, I still
> believe the assignment was a mistake. It could have been a tragic mistake.[55]

Another problem stemmed from a mismatch between doctrine and conditions on the ground in Vietnam. The development of doctrine for base defense operations to meet insurgent war threats had not been undertaken by any service.[56] Moreover, until 1968, security police training was based on concepts of defense in Air Force Manual (AFM) 207-1.[57] These concepts were primarily designed for the defense of bases located within the United States, not in a COIN

---

[51] For an overview of air base ground defense tactics and procedures, see Miranda Priebe, Alan J. Vick, Jacob L. Heim, and Meagan L. Smith, *Distributed Operations in a Contested Environment: Implications for USAF Force Presentation*, Santa Monica, Calif.: RAND Corporation, RR-2959-AF, 2019, pp. 34–42.

[52] Glen E. Christensen, *Air Base Defense in the Twenty-First Century*, Fort Leavenworth, Kan.: School of Advanced Military Studies, 2007, p. 18, notes that Operation Safe Side was the first time since World War II that the U.S. military dedicated units specifically to defending and protecting U.S. air bases. It was deemed a "crash unit" of sorts because the personnel were quickly trained and developed because of the urgency of the threat.

[53] Operation Safe Side was modeled on an RAF regiment, a light infantry battalion created for air base defense. Safe Side deployed a single squadron to Vietnam, roughly a battalion-size force, but this force was not used as originally envisioned, as a large force able to defeat a battalion-size attack. Rather, elements of the squadron, typically only a section of one officer and 38 enlisted personnel, were deployed to threatened bases. See Fox, 1979, pp. 110–112.

[54] Caudill, 2014, p. 13.

[55] USAF Inspector General SS, "Manpower Restrictions," memorandum, in *Defense of Air Bases Inspector General Report*, Vol. V, No. 1, February 17, 1968.

[56] Theodore C. Williams, "US Air Force Ground Defense System," essay, Carlisle Barracks, Pa.: U.S. Army War College, December 2, 1968, pp. 18–19. Williams further notes that "Army doctrine of the period contains virtually nothing on the defense of Army airfields much less Air Force air bases."

[57] AFM 207-1, *Doctrine and Requirements for Security of Air Force Weapons Systems*, Washington, D.C.: Department of the Air Force, June 10, 1964.

setting.[58] A 1968 USAF Inspector General memo on the defense of air bases noted: "As dramatically shown during the 1968 Tet Offensive, the stateside concept of security as required by AFM 207-1 was woefully inadequate for a guerrilla and insurgency type of war."[59] U.S. Air Force security police were also hampered by unsuccessful efforts to obtain light infantry training for their personnel. These difficulties and a dearth of doctrine contributed to C2 problems as well. The same Inspector General report bemoaned the fact that tactical security forces were commanded by wing commanders, who were often not well versed in ground defense techniques and requirements.

The Failed Memorandum of Agreement

In 1971, the Air Force attempted to attain a formal MOU with the Army on defense of air bases. As part of this effort, the Air Force asked the Army to provide sufficient forces to conduct ground operations close to bases and installations to counter any credible threats to Air Force assets and operational capabilities, to detect and counter standoff weapon attacks against air base perimeters with deterrent forces no smaller than a platoon (approximately 30 men), to provide intelligence information to air bases and installation commanders to enable perimeter defenses and readiness, and to provide liaison between external defense forces and internal Air Force base defense forces for coordination of fires, support, and related matters. The Air Force would in turn provide resources adequate for perimeter and internal defense. It would also provide forces capable of countering a platoon-size attack, gathering intelligence, and collecting human-source information in the vicinity of its bases. Finally, it would offer air strike, reconnaissance, and surveillance support to Army forces area base defense. The Army demurred.

A letter to Lt Gen Jay Robbins, Vice Commander, Tactical Air Command dated April 12, 1971, from the Assistant Vice CSAF detailed this failed initiative:

> We offered the Army a formal agreement on ground defense of overseas air bases/installations. In their response, the Army agreed that ground defense of air bases is an Army responsibility in areas where the Army maintains forces and has area defense responsibility. They assured us that the Army will fulfill this mission to the best of its capabilities. They also agreed that in other areas, responsibility for base defense would be determined on a case-by-case basis by the unified commander. The Army was not willing to make additional

---

[58] For example, they did not include infantry fundamentals, such as firing heavy weapons (mortars and .50 caliber machine guns), construction of towers and bunkers, firing at night, and establishing proper fields of fire. See Rector, 1970, pp. 25–26.

[59] USAF Inspector General SS, 1968. The Tet offensive was an inflection point for the war and for air base defense manpower needs because of the increased attacks on air bases and poor defenses. In response to the Tet offensive attacks on "Bien Hoa and Tan Son Nhut" air bases, the Air Force expressed "a requirement for 448 additional Security Police personnel, 55 M-113 armored personnel carriers and 37 armored 40mm gun carriers (Dusters)."

commitments, indicating that in their view JCS Pub 2 guidance was adequate and a detailed agreement unnecessary.[60]

The Army's insistence on the sufficiency of JCS Pub 2 was unsatisfying from the Air Force point of view. Even Army personnel conceded that, although JCS Pub 2 was unclear with respect to responsibilities for Air Force base defense, "it had generally been assumed that it is an Army responsibility inherent in a concept which relates the defense of static installations to the overall defense of land masses in which they are situated, a concept highly suspect particularly in insurgent war situations."[61] Thus, the prevailing doctrine only served to reinforce intra-service tension.

In the absence of a formal agreement, CSAF's office offered guidelines for ground base defense of overseas air bases and installations.[62] The guidelines noted that JCS Pub 2 assigned the base commander responsibility for local base defense of his command and the area commander responsibility for the overall defense of all bases in his area. The unified commander was responsible for delineating responsibilities for local defense areas. To facilitate such delineation, the Air Force offered guidance that included the following:

- The U.S. Army should provide resources for ground defense of air bases beyond the installation perimeters.
- The Air Force should provide resources for perimeter and internal security of bases.
- When Army forces are unavailable, the provision of U.S. Army resources for ground-based defense will be subject to negotiation on a case-by-case basis.

The miscarried effort essentially left the Air Force in the same position as in the late 1960s.[63]

Issue Unresolved

In summary, the Air Force made multiple attempts to improve base defenses in this new wartime environment. Unfortunately, the problem remained largely unresolved at the end of the war. The Air Force had previously taken for granted that air bases in rear areas would be secure from ground threats and that the Army would provide whatever forces were needed to protect the bases. Certainly, nothing in previous directives on roles and missions hinted that the Air Force was expected to develop its own infantry forces. Thus, it had to scramble to create ad hoc

---

[60] Assistant Vice Chief of Staff of the Air Force, letter to Tactical Air Command General Robbins, Washington, D.C.: Department of the Air Force, Office of the Chief of Staff, April 12, 1971 (Department of the Air Force Archives at Maxwell AFB, Ala.).

[61] Williams, 1968, p. 24.

[62] It is worth adding that the Air Force guidelines were, essentially, what the Air Force offered to the Army in an attempt to reach an agreement.

[63] The Air Force therefore also undertook efforts to acquire its own in-house capabilities for base defense. In 1969, it requested an Air Force owned and operated helicopter. Pacific Air Forces, "Gunship Program for Air Base Defense," Maxwell AFB, Ala.: U.S. Air Force Archives, Maxwell AFB, April 7, 1969.

organizations and train for a mission that the Air Force did not want, that was far from its core competencies, and that it saw as unique to this one conflict. Neither was the Army prepared for the demands of this war, where it faced both well-trained and well-equipped NVA mainline forces and, simultaneously, faced an enduring (typically lower-level) rear-area security problem. Neither Army doctrine nor force structure were well aligned with these dual demands.

Manpower issues affecting all services in Vietnam and MACV's emphasis on offensive operations severely limited Air Force options. Utilizing and expanding its security police proved only partially effective. Passive measures, especially the construction of 373 hardened aircraft shelters, significantly reduced losses from mortar and rocket attacks.[64] But, ultimately, the Air Force was unable to reconcile its differences with the Army, differences that were, in part, the consequences of ambiguous doctrine and differing priorities. Once the war was over, both services were happy to return to the better-understood challenges of conventional war in Europe, quickly dismissing Vietnam as a one-off scenario. Yet these problems of air base security would revisit these two services in the COIN wars of the following century.

Chapter 5 begins in Cold War Europe, where the Air Force–Army conversation shifted from the defense of air bases against insurgents to defense against Soviet and Warsaw Pact aircraft and missiles.

---

[64] This analysis of losses is from Vick, 1995, p. 70. For more on the hardened aircraft shelter program in Vietnam, see Fox, 1979, pp. 70–71, and Weitze, 2001, pp. 239–240.

5. Roles and Missions: Cold War Europe to Today

1960s and 1970s: Europe

The prevailing theme for the U.S. Air Force in the 1960s and, especially, the 1970s was modernization. The war in Vietnam was seen as somewhat of a disruption to a process of modernization, weapon acquisition, and development that began after World War II.[1] This was especially true for the Air Force in Europe. The Air Force maintained its nuclear deterrent posture on the continent, but the war in Southeast Asia had forced the Air Force to relinquish some of its fighter squadrons for that effort.[2] By the time war was winding down, the military had begun a process of reorienting its doctrine and planning for the NATO and European theaters.

For the Air Force, this meant resumption of building conventional capabilities in Europe, strengthening NATO's force posture, and increasing the survivability of forward bases and aircraft. It also meant upgrading NATO interoperability and integration, two long-standing issues. However, differences with the Army—especially over doctrine—continued to hamper Air Force efforts to consolidate its air base defenses. Difficulties notwithstanding, the 1960s and 1970s brought at least two positive developments between the two services. The first was a formal agreement between the service chiefs; the second was the assignment of Army units to defend U.S. air bases in Europe. This section briefly covers these developments.

1960s

In Europe during the 1960s, integration was a key Air Force goal. USAFE saw its air defense responsibilities as protecting U.S. forces and installations from air attack and providing combat-ready forces to NATO in the event of war.[3] However, NATO specifically reserved air defense as a national responsibility of the signatory nations. This complicated air defense responsibilities for the Air Force and encouraged the Supreme Allied Commander Europe to push for an

---

[1] Walton S. Moody and Jacob Neufeld, "Chapter 21: Modernizing After Vietnam," in Bernard C. Nalty, ed., *Winged Shield, Winged Sword: A History of the United States Air Force*, Vol. II: *1950–1997*, Washington, D.C.: U.S. Air Force, 1997.

[2] To compensate, the Air Force earmarked some U.S.-based squadrons for deployment to European bases in an emergency (Trest, 1998, p. 213).

[3] During the early 1960s, USAFE divided these responsibilities into four geographic regions: Turkey, Libya, Morocco, and West Germany. The air defense for Spain was also included in April 1960. USAFE maintained its greatest air defense capability in central Europe, where the threat from Soviet missiles and sorties was greatest. See USAFE, "Structure and Realignment of Air Defense Forces," in USAFE, *History of United States Air Forces in Europe: 1 January–30 June 1960*, Vol. 1: *Narrative*, Ramstein AB, Germany: Office of the Command Historian, 1960, p. 44.

integrated NATO air defense system that would be more capable of meeting the threat of the air weapons of the time. The Air Force had been specifically dependent on the U.S. Army for SHORAD of its bases since 1956. But importantly for the Air Force, in 1959 USAFE attained one of its long-standing objectives: The U.S. Commander-in-Chief, European Command (USCINCEUR) designated the Commander-in-Chief, USAFE (CINCUSAFE) as the theatre air defense commander with operational control over all U.S. Army Europe (USAREUR) SAMs deployed in Europe. USAFE command was also pushing for closer integration of Air Force and Army air defense weapons and systems on the continent. For instance, in August 1960, work was completed on the JCS Alerting Network, a telephone automatic switching installation at the Air Force's Headquarters Third Air Force. The system enabled establishing a worldwide voice conference network in 20 seconds with an automatic alerting system. The same year, CINCUSAFE also pushed for a NATO hardening program at its bases.[4]

In spite of such aims and improvements, deficiencies persisted. In 1960, USAFE specifically noted three: overlapping C2 channels, the obsolescence of the F-86D interceptor, and the limited track-handing potential of existing manual systems. To address these issues, the command pursued modernization efforts that included consolidating USAFE air defense resources under the 86th Air Division in late 1960, standing down its remaining F-86D squadrons for deactivation, and beginning preparations to deploy F-102s for the interceptor role. It also made efforts to improve tracking capabilities by employing a semiautomatic data handing, transmitting, and display system.[5]

However, these technical improvements did nothing to address one of the Air Force's major problems when it came to defending its European bases: doctrinal differences with the U.S. Army. The official 1960 history of USAFE made this point explicit:

> Modernization of the air defense system equipment in central Europe would not, of course, solve all of the problems in this CINCUSAFE mission; particularly, the US Air Force and US Army doctrinal differences which years of negotiation had been unable to reconcile. For example, in September 1960, USAFE and USAREUR were still trying to draft an acceptable memorandum of agreement with respect to the deployment, employment and operational control of the Hawk SAMs programmed for deploying to USAREUR beginning in 1961, although USCINCEUR had ostensibly settled the basic question of CINCUSAFE's operational control more than a year before. The respective position of the two headquarters on these matters were still poles apart in 1960.[6]

The doctrinal rifts were partially a consequence of conceptual disagreements between the Army and Air Force over how to fight the modern air defense battle. Army artillery doctrine

---

[4] USAFE, "Protection of the Forces," in USAFE, *History of United States Air Forces in Europe: 1 July–31 December 1960*, Vol. 1: *Narrative*, Ramstein AB, Germany: Office of the Command Historian, 1961, pp. 54, 67.

[5] USAFE, 1961, pp. 55–56.

[6] USAFE, 1961, p. 56.

emphasized SAMs as the primary air defense weapon, often stressing unit mobility. The Army therefore sought autonomy for its SAMs battalions—not wanting them to be subject to centralized control and direction of all air defense assets during a war.[7] The Air Force's position was that centralized C2 was essential for this mission. The passing decade would bring some mutual cooperation related to the function of air base defense but would do little to eliminate these essentially differing and entrenched points of view.

In an effort to partially address these issues, the Air Force and Army chiefs of staff signed a statement of agreed doctrine for (overseas) area air defense in July 1962. Often referred to as the LeMay-Decker Agreement,[8] the document stipulated that a coordinated and integrated air defense system under a single commander was essential to successful theater air operations and that unnecessary restrictions should not impede optimal employment of air defense systems. The agreement specified that the unified commander would normally appoint the air component commander as the area air defense commander (ADC). The ADC was fully responsible for and would have full authority over the air defense of his region—with manned interceptors, SAMs, and C2 systems (regardless of service)—integrated into a single air defense system.[9] In an important regard, the agreement was a victory for the Air Force. The ADC role in the European theater was one the Army coveted for its commander-in-chief in Europe. The LeMay-Decker Agreement was followed up by JCS Publication No. 8, *Doctrine for Air Defense from Oversea Land Areas*, which codified the tenets of the LeMay-Decker Agreement.[10]

The 1960s also saw NATO adopt its flexible response strategy, part of a broader U.S. deterrent initiative that combined diplomatic, military, and political options against the Soviets. This included a shift to improve both passive and active defense measures. The Air Force therefore began a program of dispersal and hardening. USAFE noted in the early 1960s that

---

[7] It is also worth pointing out that, in the mid-1960s, the Army was encountering its own internal debate concerning artillery. During this time, the Army possessed a single artillery branch that encompassed both field artillery and AAA. Coming out of World War II, the Army was motivated to form this combined artillery arrangement in part to prevent the newly formed Air Force from integrating the antiaircraft role into its own service remit. However, this combined artillery branch proved internally problematic for the Army. The two branches tended in different directions, with field artillery retaining its maneuver support function and antiaircraft emphasizing point defense of the U.S. and its NATO allies. Difficulties in training across the functions manifested themselves in Korea and Vietnam, and officers struggled with cross assignments. This added to the internal debate about separating the two artilleries. The Army finally made the separation official with its General Order 25 on June 14, 1968, establishing Air Defense Artillery as a basic branch of the Army. See Hamilton, 2009, pp. 195–202.

[8] The agreement is reproduced in Wolf, 1987, pp. 216–218.

[9] Wolf, 1987, pp. 7–8, offers an assessment of the agreement's importance. He notes that the Vandenberg-Collins Agreement had embodied the essentiality of a single commander to coordinate the joint efforts between the two services for air defense. However, after Secretary Wilson's 1956 memorandum on roles and missions, the two services differed on whether an Army field commander would be responsible for air defense and control air operations over his combat area. Ultimately, the overseas unified commanders were able to resolve the issues solidified in the LeMay-Decker Agreement.

[10] Robert F. Futrell, *Ideas, Concepts, Doctrine: Basic Thinking in the United States Air Force*, Vol. II: *1961–1984*, Maxwell AFB, Ala.: Air University Press, 1989, p. 192.

hardening alone would not result in an adequately protected strike force but was a necessary start.[11] In late 1960, Headquarters USAFE requested and was granted permission to implement a hardening program through USCINCEUR and NATO channels. The program would continue into the next decade. The Supreme Allied Commander Europe would later require submission of a general and detailed programs for physical protection of airfields for each eligible airfield to qualify for NATO common funding. General programs included such elements as runway repair, aircraft dispersal, and protective aircraft shelters. By June 1972, U.S. general programs had been submitted for all eligible USAFE bases except those in the UK. By mid-1972, the USAFE aircraft shelter program had completed 360 shelters across eight European bases and had funded an additional 54.[12]

Vulcan and Chaparral

Point air defense in Europe was increasingly a cause of concern in the late 1960s and into the 1970s because of the growing Soviet air threat. As early as 1960, USAFE had estimated that the Soviets had roughly 10,000 aircraft of various types within striking distance of targets within the USAFE Central Europe area of operation.[13] However, by the middle of the 1960s, it was clear to the Army that its mobile air defense weapon systems (the M42 "Duster" 40-mm and M45 Quad .50 caliber machine gun) were obsolete. As a solution to the problem, the Army turned to the Vulcan gun and Chaparral missile. The Vulcan was a 20-mm Gatling–type gun (adapted from the Air Force M-61 rotary cannon) that could be mounted on an armored personnel vehicle or towed. The Chaparral missile, developed in the mid-1960s, was an adaptation of the Navy's heat-seeking Sidewinder missile. Designed to defeat high-speed fighter aircraft, it consisted of four missiles mounted on launch rails, typically mounted on a full-tracked vehicle. An Army Chaparral-Vulcan battalion consisted of two Chaparral missile batteries and two Vulcan gun batteries.[14]

In 1968, the first set of Vulcans was sent to Vietnam after the Air Defense Board conducted tests to ensure the weapon was ready for combat.[15] The test unit in Vietnam was used to perform perimeter security and other functions around the base.[16] By 1970, the Army had activated eight Chaparral-Vulcan battalions that were to become part of the air defense system in Europe. As

---

[11] For this reason, USAFE also supported the introduction into the command's weapon inventory of both aircraft and missiles "that could be truly mobile" (USAFE, 1961, p. 60).

[12] USAFE, "USAFE Shelter Construction Status," in USAFE, *History of United States Air Forces in Europe: For Fiscal Year 1972*, Vol. 1: *Narrative*, Ramstein AB, Germany: Office of the Command Historian, 1973, pp. 56–57. For more on the USAFE shelter program, see Benson, 1981.

[13] USAFE, 1961, p. 58.

[14] Hamilton, 2009, pp. 218, 222.

[15] Hamilton, 2009, p. 223.

[16] Hamilton, 2009, p. 223.

noted earlier, advances in Soviet weaponry presented an acute threat to bases. But NATO considered point air defense a national responsibility. Therefore, collocated operational bases were to be protected by host countries. Point air defense for USAFE's primary bases in Europe was a responsibility of the U.S. Army. In 1969, the Army undertook a program to streamline air defense assets in Europe. As part of this program, Army units—two Chaparral-Vulcan battalions of the 32nd Army Air Defense Command—were assigned to defend five USAFE air bases and the U.S. Army's Kaiserslautern logistical complex in West Germany.[17] The arrangement would persist during the following decades.

This agreement, in which point air defense of air bases was the primary mission for some Army air defense artillery units, was perhaps a high point in Army–Air Force cooperation. Unfortunately, there were far more Air Force MOBs in Europe than Army artillery battalions to defend them. The 1972 USAFE official history noted that two of the six USAREUR Chaparral-Vulcan battalions stationed in Central Europe were assigned to defend five of its seven MOBs in Germany.[18] However, the same document lamented the lack of active air defense forces to protect the command's bases in Turkey, Italy, the Netherlands, and the UK.[19]

The commander of the 108th Air Defense Artillery Brigade in Germany observed in 1984 that the 108th was the only air defense brigade in the entire Army that had air base defense as its primary mission, a reflection of the Army's commitment to the function in the European theater. However, the commander also noted that the Army had yet to develop tactical doctrine to assist air defense artillery brigades in performing this mission.[20] C2 issues were an additional concern. While the Army air defense commander had responsibility for command and design of the defense, the authority to issue weapon control orders rested with the supported Air Force wing commander. In spite of these concerns, this joint arrangement served USAFE's greater interests and may also have fulfilled a deterrence role. By 1983, USAFE had established a working group with its headquarters and the 108th Air Defense Artillery Brigade, which "proved effective in opening communications between the Air Force and the Army Air Defense" and emphasized the Chaparral and Vulcan systems.[21]

The late 1970s and early 1980s marked a period of effective collaboration between the Army and Air Force. In 1976, the commanders-in-chief of the USAREUR and USAFE set up the Joint Actions Steering Committee and the Directorate of Air-Land Forces Application (DALFA) to

---

[17] Walter Elkins, "32nd Army Air Defense Command," U.S. Army in Germany website, undated.

[18] The MOBs in Germany at that time were Bitburg, Hahn, Rhein-Main, Spangdahlem, Sembach, Zweibrucke, and Ramstein air bases. See Benson, 1981, pp. 155–158.

[19] USAFE, 1973, p. 57.

[20] Rocco, 1984.

[21] USAFE, *History of United States Air Forces in Europe: Calendar Year 1983*, Vol. 1: *Narrative*, Ramstein AB, Germany: Office of the Command Historian, 1984, p. 354.

undertake projects toward improving readiness.[22] By the end of 1977, the Joint Actions Steering Committee had undertaken five projects, one of which included the point air defense of air bases. For its part, the Air Force certainly wanted the Army to improve and expand its air defense systems, and these joint committees were a way of effectively communicating concerns. This era was a high point in cooperation and communication between the services and laid fertile ground for the establishment of MOUs that would follow in the 1980s.

1980s: Period of Army–Air Force Cooperation and Frustration

After Vietnam, the United States pivoted to focus on defending against a Soviet invasion of Europe but lacked the resources to do so effectively. DoD once again found itself in a postconflict period of budget contractions. Following a prolonged, costly, and unpopular war, the U.S. public had little appetite for defense spending.[23] This context forced the services to again address trade-offs and make difficult decisions about which systems they deemed had the highest priority and where they could accept risks. It also motivated the Army and Air Force to seek coordination as the service leaders faced looming budget cuts.

The choices about which capabilities to emphasize led to both frustration and collaboration from the Air Force and Army, which had implications for air base defense. What came to be known as the *31 Initiatives* marked a period of remarkable collaboration between the two services. These successes notwithstanding, the era was also one of discord. The Army experienced highly public difficulties with its SHORAD modernization programs because it was forced to cancel two costly efforts, the division air defense gun (Sergeant York) and the Roland missile. These hitches both frustrated the Air Force and negatively affected air base defense. The next subsections explore these two inflection points in the 1980s, which reveal the changing nature (and priority) of AMD roles and missions in the post-Vietnam period.

31 Initiatives

In the shadow of a growing Soviet and Warsaw Pact threat, the U.S. military developed new concepts to defeat the Soviet Union. This threat and two conflicts precipitated the Army's AirLand Battle doctrine.[24] The first was the U.S. experience in Vietnam, a decade-long irregular

---

[22] USAFE, *History of United States Air Forces in Europe: Calendar Year 1977*, Vol. 1: *Narrative*, Ramstein AB, Germany: Office of the Command Historian, 1978, p. 162. The USAEUR/USAFE DALFA initiative was modeled on the Air-Land Forces Application Center (ALFA) established in 1975 by the commanders of the USAF Tactical Air Command and Army Training and Doctrine Command and located at Langley AFB, Virginia. In 1992, ALFA became the Air Land Sea Application Center (ALSA). For more on the history of ALFA/ALSA, see Air Land Sea Application Center, "ALSA Roadshow," briefing slides, September 5, 2019.

[23] Manfred R., *The Conventional Arms Balance, Part 3: Deterring Nuclear War in Europe*, Washington, D.C.: The Heritage Foundation, July 16, 1986.

[24] John L. Romjue, *From Active Defense to AirLand Battle: The Development of Army Doctrine 1973–1982*, Fort Monroe, Va.: Historical Office, U.S. Army Training and Doctrine Command, June 1984.

war that deemphasized conventional combat strengths. The second was the 1973 Yom Kippur War, when Arab forces, armed with Soviet equipment and following Soviet doctrine, put the state of Israel and its vaunted defense forces in peril.[25] The Soviets had modernized their offensive capabilities while the United States was bogged down fighting a jungle war in Southeast Asia. This sudden realization spurred the Army and the Air Force to meet the challenge of the defense of NATO. One result was the publication of a new Army Field Manual (FM) 100-5, *Operations*.[26] FM 100-5 represented the capstone operational doctrine of the Army: the AirLand Battle, a fighting concept that emphasized close integration between land and air forces.[27] The doctrine envisioned a widening of the battlespace to include both the "deep" and "close" battle against a large and technologically sophisticated adversary, such as the Soviet Union. Developed in the 1970s, AirLand Battle would replace the earlier Active Defense doctrine, promulgated by the Army in 1976.

In parallel with the development of the Army's new doctrine and to address the shared Army and Air Force problem of defending Western Europe from Soviet aggression, the Army and Air Force chiefs of staff, GEN John A. Wickham, Jr., and Gen Charles A. Gabriel, embarked on a period of intense cooperation. The results of their efforts are commonly referred to as the *31 Initiatives*.[28] On May 22, 1984, Gabriel and Wickham signed a memorandum of agreement outlining 31 initiatives to "further Air Force–Army cooperation on the battlefield."[29] The wide-ranging agreement began by acknowledging that, to fulfill their respective roles in national security objectives and defense, the Army and Air Force "must organize, train, and equip a comparable, complementary, and affordable Total Force" to maximize joint combat capability to execute "airland combat operations."[30]

The 31 Initiatives touched on seven key aspects of air land combat, ranging from air defense to the fusion of combat information.[31] Table 5.1 provides a snapshot of the initiatives that pertain to air base defense. In the following subsections, we briefly discuss Initiative 1, the 1984 MOU on air base air defense, and Joint Service Agreement (JSA) 8. (Initiative 2 is discussed under the 1984 MOU; similarly, Initiative 8 is discussed with JSA 8.)

---

[25] David E. Johnson, *Shared Problems: The Lessons of AirLand Battle and the 31 Initiatives for Multi-Domain Battle*, Santa Monica, Calif.: RAND Corporation, PE-301-A/AF, 2018, p. 2.

[26] FM 100-5, *Operations*, Washington, D.C.: Headquarters, Department of the Army, 1982.

[27] Richard W. Stewart, *American Military History*, Vol. II, *The United States Army in a Global Era, 1917–2008*, Washington, D.C.: Center of Military History, U.S. Army, 2010, p. 383.

[28] Davis, 1987, p. v; Johnson, 2018.

[29] Davis, 1987, p. 1.

[30] Wolf, 1987, p. 415. Many tout Gabriel and Wickham's personal friendship and long history of working together on a joint level as instrumental to the development of the initiatives. See Davis, 1987, p. 36.

[31] Davis, 1987, pp. v, 2, 47.

Table 5.1. 1980s Army–Air Force Initiatives Pertaining to Air Base Defense

Document	Date	Title	Category
Initiative 1	May 22, 1984	Area Surface-to-Air Missiles/Air Defense Fighter	Increase joint action and cooperation for combat, doctrine, function
Initiative 2	May 22, 1984	Point Air Defense	Increase joint action and cooperation for combat, doctrine, function
Initiative 8	May 22, 1984	Air Base Ground Defense	Clarify roles and missions
MOU on Air Base Air Defense	July 13, 1984	Memorandum of Understanding on United States Army (USA)/United States Air Force (USAF) Responsibilities for Air Base Air Defense	Clarify roles and missions
JSA 8 on Ground Defense of Air Bases	April 25, 1985	Joint Service Agreement on United States Army (USA)/United States Air Force (USAF) Agreement for the Ground Defense of Air Force Bases and Installations	Clarify roles and missions

SOURCE: Information from Davis, 1987, p. 2.

Initiative 1—Area Surface-to-Air Missiles/Air Defense Fighter

The first initiative made three specific recommendations to improve cooperation on point air defense. First, the Air Force should become involved in the requirement and development phases for SAMs, previously dominated by the Army. Second, the Air Force and Army should jointly determine the optimal combination of area SAMs and air defense fighters.[32] Additionally, the first initiative served as a catalyst for other joint SAM endeavors. Initiative 1 led to a joint study on area SAMs that was aimed at assessing the benefits and drawbacks of transferring Patriot from the Army to the Air Force. This study touched on a larger issue about the consolidation or decentralization of area defense. The broader disputes about who should control which air defense missions continued between the Army and Air Force because their efforts overlapped within the DCA mission. Ultimately, the study advised the services to refrain from transferring SAMs from the Army to the Air Force. The joint working group noted that, while it would be feasible to transfer Patriot to the Air Force, this would not provide enough operational or financial benefits to warrant the organizational overhaul.[33]

Memorandum of Understanding on Air Base Air Defense

The MOU on air base defense was the product of Initiative 2 on point air defense.[34] Signed July 13, 1984, the MOU stated that "Air base air defense at USAF bases is a joint responsibility of the US Army and Air Force. To this end, the Air Force will be responsible for submitting

---

[32] Davis, 1987, p. 49.

[33] Mark A. Robershotte and Greg H. Parlier, "Army Retains Patriot," *Air Defense Artillery*, Spring 1986.

[34] Davis, 1987, p. 49.

requirements for air base air defense to the Army for support."[35] The Army will be "responsible for ground-based air defense at Air Force . . . MOBs . . . worldwide," which is to say that the Army would provide necessary ground-based SHORAD systems, whether guns or missiles.[36] The Air Force agreed to support "the Army's efforts to obtain additional force structure and funding to . . . address specific shortfalls."[37] Most important, the memo stated that "[i]f the Army is unable to provide adequate support, then the Air Force may pursue alternative solutions such as cooperative arrangements with host nations or deployment of USAF organic point air defense capability."[38] This is the most striking element in the MOU, giving the Air Force the option of developing its own SHORAD systems in the event the Army failed to deliver. Were the MOU still in place, it would afford a foundation for the Air Force to develop organic cruise missile defenses. The MOU is no longer active because it stipulated a biennial review that, if missed twice, would render the MOU void.[39] While the agreement was in effect, however, the Air Force did not undertake to procure organic point air defenses; had it done so, it might have encountered procurement difficulties or been unable to substantiate that the operational or financial benefits merited such an organizational overhaul.

Joint Service Agreement on Ground Defense of Air Bases and Installations

Initiative 8 addressed Army and Air Force responsibilities for air base ground defense, providing the foundation for a JSA.[40] General Wickham and General Gabriel signed JSA 8 on April 25, 1985, to clarify responsibilities for defense of air bases from ground attack.[41] In doing so, the JSA built on tenets initially codified in the MOU from 1984 on air base air defense. This JSA provided a list of definitions for such terms as *air base ground defense*, *base or installation boundary*, and *base defense*. These definitions provided clarity on several terms important to Air Force–Army cooperation in the defense of air bases.

JSA 8 states that the Air Force will retain C2 of ground defense of air bases, with Army units reporting to the installation commander. The Army remained responsible for providing forces for any air base ground defense "operations outside designated Air Force base or installation

---

[35] Davis, 1987, p. 123.

[36] Davis, 1987, p. 123.

[37] Davis, 1987, p. 123.

[38] This discussion is based on Department of the Army and Department of the Air Force, "Memorandum of Understanding on United States Army (USA)/United States Air Force (USAF) Responsibilities for Air Base Air Defense," Washington, D.C.: Headquarters, U.S. Army and Headquarters, U.S. Air Force, July 13, 1984, as reproduced in Davis, 1987, pp. 120–124.

[39] It is likely that this provision in the agreement ultimately brought about its "unofficial" end some years later.

[40] Davis, 1987, pp. 51–53.

[41] Department of the Army and Department of the Air Force, "Joint Service Agreement on United States Army (USA)/United States Air Force (USAF) Agreement for the Ground Defense of Air Force Bases and Installations," Washington, D.C.: Headquarters, U.S. Army and Headquarters, U.S. Air Force, April 25, 1985, as reproduced in Davis, 1987, pp. 125–131.

boundaries," while the Air Force would provide resources for physical security and internal defense within the perimeter of bases and installations.[42] The services agreed that the objective of air base ground defense, and the JSA, was to allow the Air Force to generate combat power from fixed facilities and ensure "sortie generation."[43]

In fact, the 1984 and 1985 agreements and the new doctrinal scheme they laid out for base defense produced some optimism at the Air Force's European headquarters. The 1985 official history of USAFE stated that, in support of the agreements, the Army planned to organize and train a force of soldiers for the base security role. It further noted that, "[b]eginning in FY 86, the Army would start training approximately 7,500 U.S. Air Force security police annually in infantry tactics through a modified basic infantry training course."[44] Such a training program would ensure that Army and Air Force security forces operating near the air bases were trained in the same tactics and doctrine. While it is unclear whether such training took place, the comment is indicative of a mutual approach to defending air bases in light of the agreements. Unfortunately, this spirit of cooperation would not last.

Army SHORAD Modernization Shortfalls

Acquisition debacles, interservice feuds, and public criticism plagued the Army's modernization efforts in the 1980s. The service could not field or produce a viable SHORAD system. Consequently, the Air Force grew frustrated because it felt its air bases were vulnerable to the Soviet threat, and the Army had not made ground-based air defense and SHORAD systems a priority. Because the Air Force and Army shared the air defense mission, certain systems and functions fell under the Army, leaving the Air Force optionless when it came to acquisition and procurement. The Army had responsibility for ground-based point defenses (SAMs and AAA), while the Air Force had ownership of air-to-air defense.[45]

The Army began to modernize its interim SHORAD capabilities in earnest after it became evident in the 1960s that the Chaparral missile and the Vulcan gun could not perform in all weather conditions. Observations from the 1973 Arab-Israeli war showcased the adeptness of Soviet missile and gun systems and highlighted the lack of a comparable U.S. air defense system. The Army had not fielded a new gun system since the Vulcan. In light of these factors, the Army concluded that it needed to fill this close-air-defense gap.[46]

---

[42] Davis, 1987, p. 128.

[43] Harold R. Winton, "An Ambivalent Partnership: US Army and Air Force Perspectives on Air-Ground Operations, 1973–1990," in Phillip S. Meilinger, ed., *The Paths of Heaven: The Evolution of Airpower Theory*, Maxwell AFB, Ala.: Air University Press, 1997, p. 430.

[44] USAFE, *History of United States Air Forces in Europe: Calendar Year 1985*, Vol. 1: *Narrative*, Ramstein AB, Germany: Office of the Command Historian, 1986, p. 330.

[45] John T. Correll, "Air Defense from the Ground Up," *Air Force Magazine*, Vol. 6, No. 7. 1983, pp. 37–43.

[46] Hamilton, 2009, p. 259.

After years of tests, studies, and discussions, the Army decided to pursue the Sergeant York gun and the Franco-German Roland missile as replacement systems for air defense.[47] While the Army hoped these systems would prove effective, they failed for performance, cost, and bureaucratic reasons. In 1981, the Army canceled the Roland missile program (designed to replace the Chaparral missile) and, in 1985, terminated the Sergeant York gun (designed to replace the Vulcan gun). Both weapons were designed for SHORAD functions, which included ground-based air base defense. These cancellations and the manner in which they were carried out strained relations with the Air Force. The Army offered no advance warning to the Air Force of its plan to cancel the U.S. Roland in 1981. The Air Force learned of the system's cancellation only by reviewing the Army's budget submission to Congress.[48] The lack of coordination was a direct reflection of service tensions.

For its part, the Army was also frustrated; it had made ground-based air defense and SHORAD priorities, as demonstrated with the Sergeant York and Roland—but just could not produce the systems.[49] The Army did not find a SHORAD solution until it settled on the short-range Avenger air defense system in the 1990s. The next sections explore the Sergeant York gun and Roland missile modernization mishaps in greater detail.

Sergeant York Gun

In post-Vietnam America, the services were forced to operate in an often tense, budget-constrained environment.[50] While the U.S. military was finding ways to efficiently use funds and cut systems, the Soviets were modernizing and improving their capabilities. The Soviet air threat again became a concern. The U.S. military desperately needed to improve its capability to combat Soviet advances. The solution became a series of efforts, one of which included the Sergeant York gun, which was born at the Air Defense School in Texas.[51]

Before pursuing the Sergeant York, the Army laid out the technical criteria it sought in a new gun system. The system had to be self-propelled, able to keep pace with mechanized battalions, and able to target and defeat highly agile fighter aircraft and helicopters. The Sergeant York met many of these system-specific requirements. The challenge, however, stemmed primarily from the weapon system's inability to pass operational testing. Its tracking radar was a particular

---

[47] Russell A. Hinds, *The Avenger and SGT York: An Examination of Two Air Defense Systems Non-Developmental Item Acquisition Programs*, thesis, Monterey, Calif.: Naval Postgraduate School, March 1995, p. 5. See also McNaugher, 1989, pp. 103–104.

[48] Correll, 1983.

[49] On this point, see O. B. Koropey, *It Seemed Like a Great Idea at the Time: The Story of the Sergeant York Air Defense Gun*, Redstone Arsenal, Ala.: Historical Office, U.S. Army Materiel Command, January 1, 1993.

[50] It is worth also adding that, in addition to the end of the Vietnam War, the Nixon Doctrine changed the U.S. strategy from 2.5 to 1.5 wars and abjured involvement in contesting ongoing and future wars of national liberation.

[51] Hinds, 1995, pp. 5–8. In particular, MG C. J. Levan at the Air Defense Artillery School in Fort Bliss, Texas, took ownership of the Sergeant York.

problem: It had difficulty differentiating helicopters from trees.[52] In spite of these difficulties, the Army successfully accepted the first Sergeant York firing unit on March 13, 1984.[53] The Army requested 132 guns for FY 1985 and 144 guns for FY 1986. The Army justified these requests by stating that the guns would be used for air defense to protect against armed helicopters and fixed-wing aircraft.[54] These hopeful budget requests were soon to become moot because the Sergeant York program was canceled.

Several factors contributed to the Sergeant York's eventual downfall. From its inception, the Sergeant York faced an uphill acquisition process full of public criticism, poor technical performance, and outcries over high costs. First, the technical complexity of the gun escalated its overall cost. To be an all-weather system, the Sergeant York required an advanced radar (derived from the F-16 fighter) and automated fire control, both costly and complex technologies. In light of the system's rising price tag, many found its technical implementation underwhelming. Specifically, the system did not perform well against helicopters and airplanes. Second, the system was on an accelerated acquisition schedule. The haste only exacerbated preexisting issues.[55]

Ultimately, the Sergeant York was canceled because it was deemed unaffordable.[56] After the Pentagon had spent ten years and $1.8 billion on the Sergeant York, Secretary of Defense Caspar W. Weinberger canceled the system in August 1985.[57] Weinberger argued that the system was simply not worth its cost. When the weapon was canceled, 65 of the 618 ordered guns had already been delivered.[58]

The cancellation was a debacle for the Army air defense community. A *New York Times* article published shortly after the cancellation noted that, "[i]f a weapon can be said to have died

---

[52] For more on these and other problems related to testing, see Russell Phillips, "An Ineffective System: The M247 Sergeant York," Russell Phillips Military Technology and History website, undated.

[53] Hamilton, 2009, p. 263.

[54] Department of the Army, *Justification of Estimates for Fiscal Year 1985, Submitted to Congress February 1984, Procurement Programs, Aircraft, Missiles, Weapons, and Tracked Combat Vehicles, Ammunition, other Procurement, and Construction Programs*, Washington, D.C.: Office of the Deputy Chief of Staff for Research, Development, and Acquisition, February 1984.

[55] "Demise of the SGT. York: Model Turns to Dud," *New York Times*, December 2, 1985.

[56] The Army dealt with competition coming from the Abrams tank and problems acquiring the correct chassis for the gun. Additionally, the cost of the Sergeant York far exceeded that of the Abrams. One Sergeant York cost $6.6 million, but a single Abrams tank cost $2.8 million. Perhaps illustrating the system's steep costs, acquisition of the Sergeant York occurred over a series of years. See Hamilton, 2009, p. 260.

[57] John S. DeMott, "Son of the Sergeant York," *Time*, June 24, 2001; Rudy Abramson, "Weinberger Kills Anti-Aircraft Gun: After $1.8 Billion, He Says Sgt. York Is Ineffective, Not Worth Further Cost," *Los Angeles Times*, August 28, 1985.

[58] Abramson, 1985.

of embarrassment, the antiaircraft gun called Sergeant York did."[59] Tactically, the Sergeant York cancellation left a large air defense gap in terms of protecting against Soviet aircraft and helicopters. While the aging Chaparral and Vulcans were still in the Army, their operational limitations (daytime use and only in fair weather) meant that both Army maneuver forces and fixed facilities were increasingly vulnerable to enemy attack. The quality of NATO air forces combined with the advanced Patriot air defense system somewhat reduced the need for SHORAD systems. If the Cold War had continued, the Army likely would have once again pursued more-advanced SHORAD systems but instead found that the Avenger air defense system mounted on a humvee and equipped with Stinger missiles and a .50 caliber heavy machine gun was adequate for the early post–Cold War environment.[60]

Roland Missile

The Army spent most of the late 1960s and 1970s searching for a viable SHORAD missile system. The all-weather Franco-German Roland missile system became part of this story in 1975. On January 9, 1975, the Army chose the Roland missile only to see it fail because of a combination of high production costs, other Army priorities for air defense funds (i.e., Patriot and the Stinger man-portable air-defense systems), and a restrictive budgetary environment.[61]

The Army chose to adopt the Roland, which was already in production in Europe, to save money and bypass the time-intensive processes involved in building its own capability. The Roland launcher was designed to fit on top of a tracked or wheeled vehicle and carried a total of ten missiles.[62] This SHORAD system was further envisioned as a complement to the Hawk, a radar-directed missile used to protect rear or forward areas from high-altitude aircraft.[63] The Roland was employed to counter low-flying aircraft in all weather conditions and for rear-area defense.[64]

Importing a system already in use was somewhat seamless for the Army. The trouble began when Army officials decided to "Americanize" the Roland—that is, develop in-house (U.S.) production processes. The Army argued that developing a U.S. manufacturing and production

---

[59] "Demise of the SGT. York...," 1985. These critiques reached the public domain as media outlets and senior military leaders publicly debated their validity, exacerbating the Army's difficulty in pursuing the Sergeant York. One senior Army leader, General Maloney, refuted arguments that the Sergeant York's cost far exceeded that of the Abrams tank, suggesting that a small number of Sergeant York systems would protect a large number of tanks. See Hamilton, 2009, p. 268.

[60] Avenger was first produced in the late 1980s ("Avenger Low Level Air Defence System, USA," webpage, Army Technology, undated).

[61] "U.S. Army Selects European Missile," *New York Times*, January 10, 1975; Fred Hiatt, "Improved Missile Too Costly for Pentagon," *Washington Post*, August 28, 1984.

[62] Hamilton, 2009, p. 260.

[63] "U.S. Army Selects ...," 1975.

[64] Comptroller General of the United States, *Evaluation of the Decision to Begin Production of the Roland Missile System*, Washington, D.C.: General Accounting Office, August 17, 1979, p. 1.

base would avoid the risk that Europe would be unable to export these systems in a wartime environment. This in-house approach ultimately contributed to the missile's termination. To create a U.S. manufacturing base for a foreign system, the Army selected two prime contractors, Boeing and Hughes Aircraft Company. The contractors were forced to translate drawings, manuals, and documents from French and German into English. This proved a costly and time-consuming endeavor.[65]

Additionally, the Army made a series of decisions that increased the cost of the missile's Americanization process. The Army elected to mount the missile on a U.S. armored vehicle chassis. It further sought a system resistant to chemical, biological, and nuclear attacks and one equipped with computer chips up to military specifications. It also wanted the weapon enhanced with electronic countermeasures to defend against EW attacks. These measures added to the growing cost of the Roland. On top of these production and technical factors, the Roland and SHORAD had low priority for the Army (the lowest priority of all air defense programs). Low prioritization was a vulnerability in an era of restricted budgets.

By 1983, MG James P. Maloney, director of Army weapon systems, determined that the Chaparral was a suitable alternative (in light of budgetary considerations) and canceled the larger Roland program.[66] With over 600 Rolands left in production, the Army chose to transfer them to the National Guard. At this point, the Army had invested more than $1 billion in a system it thought would provide SHORAD capabilities for nearly a decade. This same pattern continued when the Roland systems were sent to the National Guard. By 1985, a National Guard Roland battalion was declared ready for combat; by 1988, the same battalion was inactivated.[67]

The continued failures in SHORAD systems for point defense frustrated the Air Force in the 1980s.[68] However, this sentiment shifted in the 1990s. From 1990 to 2014, neither the Air Force nor the Army saw an urgent requirement for improvements in point defenses because low-intensity conflicts and irregular war dominated the threat landscape. This shift reflected a broader pattern within DoD during this period. Many programs were canceled or curtailed after the Cold War, particularly given the demands that arose after September 11, 2001. For example, the Air Force reduced its F-22 and B-2 buys, and the Army nixed the Crusader, its next-generation self-propelled howitzer, and the Comanche helicopter project.

---

[65] Fred Hiatt, "Improved Missile Too Costly for Pentagon," *Washington Post*, August 28, 1984.

[66] Hiatt, 1984.

[67] Hamilton, 2009, pp. 257–263.

[68] One debate during this period of frustration considered transferring primary responsibility for the Patriot air defense missile system from the Army to the Air Force. The two service chiefs—based on a joint Army–Air Force study—finally decided in 1986 against the transfer, citing possible disruptions of system financing and manning and deployment concerns. See Robershotte and Parlier, 1986.

1990s: Little Air Force or Army Interest in Air Base Air Defense

In the 1990s, with the end of the Cold War and success in the first Gulf War, the U.S. military experienced a new sense that rear areas were sanctuaries. This newfound sense of sanctuary diminished the military's focus on protecting these areas. The Air Force's behavior, rhetoric, and proposals throughout the 1990s did not stress the need for air base defense because the military had become accustomed to rear-area sanctuary. Three key operational advantages stemmed from rear-area sanctuary that allowed the services to focus on other operational needs and capabilities: the quick and effective use of airfields and ports, the ability to use rear-area MOBs to house large forces (capabilities, systems, and personnel), and the reduced lag time between forces arriving on base and leaving on subsequent missions.[69]

A couple of factors likely contributed to the Air Force and other services' pursuit of capabilities that were more offensive and to downplay the defense of air bases in the early 1990s. Operational incentives and the absence of a threat to air bases led the Air Force to pursue more-compelling demands. The slow dissolution of JSA 9 in the post–Cold War era is a reflection of such disinterest. Under the arrangements from 1985, the Army had agreed to train airmen in air base ground defense, even going so far as to stand up a school for that purpose. However, by the 1990s, and consistent with the post–Cold War environment, defense of air bases from ground attack was not a priority concern for either Air Force or Army leaders.[70] In August 1995, "Air Base Defense Training moved back to Camp Bullis, Texas, and was once again under control of the Air Force."[71]

As operational and threat factors changed, so too did the military. Similar to the 1945–1949 period, the 1990s marked an era when the services once again had to redefine their purpose, mission, and place in the larger U.S. national security apparatus.[72] With the collapse of the Soviet Union and the dissolution of the Warsaw pact, the U.S. military found itself bereft of its decades-long justification for military resources, capabilities, and overseas presence.[73] This played out in the 1990s in a congressionally mandated review of roles and missions and ongoing Air Force–Army disputes over C2 of airpower in combat theaters.

Ballistic Missile Defense in the Gulf War

During the first Gulf War, the Army deployed three AMD systems: Avenger, Patriot, and Hawk. The Patriot, an all-weather, all-altitude, long-range air defense system, was used in

---

[69] Vick, 2015a, pp. 13–14.

[70] Christensen, 2007, pp. 19–20.

[71] AFM 31-201v1, *Security Forces History*, Washington, D.C.: Department of the Air Force, August 9, 2010, p. 12.

[72] Paula Thornhill, *Demystifying the American Military: Institutions Evolution and Challenges Since 1789*, Annapolis, Md.: Naval Institute Press, 2019.

[73] Stewart, 2010, pp. 407–408.

Operation Desert Storm primarily for its anti–ballistic missile function.[74] Similarly, the Hawk, a surface-to-air system for medium-altitude threats, was used to provide air defense in Operation Desert Storm. Lacking weapons that could inflict significant damage, Saddam Hussein employed Scud tactical ballistic missiles against ports and civilian areas in Israel and Saudi Arabia during the war.[75] His gamble was that the shock of missile attacks on rear areas and population centers might break the coalition or draw Israel into the conflict.[76] To defend against the Scud attacks, the Army deployed the Patriot air defense system.[77] The weapon system, untested in combat prior to this war, played a large role psychologically, if not operationally, in lessening the strategic effects of the attacks.[78] In total, the Army deployed 21 Patriot batteries to Saudi Arabia, four batteries to Turkey, and seven batteries to Israel.[79]

In Saudi Arabia and Israel, Patriots were used for point defense of air bases and area defense of larger cities.[80] In Turkey, the Army used the Patriots solely to defend air bases located there, but Patriots were never involved in any engagements.[81] Additionally, the Army deployed Hawk and Avenger SAMs for air defense.[82] The combat debut of Avenger was deemed a success, prompting the Army to purchase an additional 679 multipurpose vehicles for its Avenger system in 1992, bringing the total Avengers in production to 1,004. Roughly 800 of those remained in service as of this writing.[83]

The Grand Bargain

By 1993, Gen Merrill McPeak, CSAF, saw AMD as rife with overlap, inefficiency, and incorrect allocation of resources. In his memoir, McPeak argued that most established militaries place air defense capabilities under the air force rather than the Army, suggesting the U.S. military should follow suit.[84] This logic prompted him to present a "grand bargain" to his Army

---

[74] "Patriot Missile Long-Range Air-Defence System," Army Technology website, undated.

[75] DoD, *Conduct of the Persian Gulf War: Final Report to Congress*, Washington, D.C., April 1992, p. 169.

[76] Thomas A. Keaney and Eliot A. Cohen, *Gulf War Air Power Survey Summary Report*, Washington, D.C.: U.S. Department of the Air Force, 1993, p. 166.

[77] Robert H. Scales, *Certain Victory: The U.S. Army in the Gulf War*, Sterling, Va.: Potomac Books, Inc., 1993, pp. 181–183.

[78] Air Defense Artillery Association, "1991 Air Defense Artillery Yearbook," *Air Defense Artillery*, Spring 1991.

[79] Keaney and Cohen, 1993, p. 756.

[80] DoD, 1992, p. 756.

[81] Keaney and Cohen, 1993, p. 756.

[82] Frank N. Schubert and Theresa L. Kraus, *The Whirlwind War: The United States Army in Operations Desert Shield and Desert Storm*, Washington, D.C.: U.S. Government Printing Office, 1995; and Air Defense Artillery Association, 1991.

[83] "Avenger Low Level . . . ," undated.

[84] Merrill A. McPeak, *Roles and Missions*, Lake Oswego, Oreg.: Lost Wingman Press, 2017, p. 311. Our review of 2012 data in *Jane's World Air Forces* suggests that McPeak overstated his case. As of 2012, 78 percent of the

counterpart, GEN Gordon Sullivan, CSA, in 1993. He offered to transfer A-10 aircraft and the close air support (CAS) mission to the Army in return for ownership of theater AMD. McPeak did not wish to strip the Army of air defenses for mobile forces but did see a need to transfer high- to medium-altitude AMD systems primarily used for theater air defense and the protection of fixed facilities in rear areas.[85] If accepted, McPeak's proposal would have resolved two enduring R&F disputes.

First, it would have freed the Air Force from persistent Army complaints about the provision of CAS. Second, it would have given the Air Force the control over theater AMD that it had long sought.

General Sullivan rejected the proposition outright, offering three objections. First, he claimed that CAS was not a traditional function of the Army; therefore, it would be an entirely new area for the service. He held fast to Army doctrine that attack helicopters do not perform the CAS mission.[86] McPeak refuted this point, stating that "what attack helicopters do walks like CAS and talks like CAS; I've done a bit of CAS myself, and it sure looks like CAS to me." Sullivan also pointed to the high cost of A-10 aircraft as too steep for the Army. McPeak countered this point as well, offering to cover the funding of these aircraft if the Army would transfer Hawk and Patriot battalions in return. Last, Sullivan argued that owning A-10 aircraft would tie the Army to air bases. Nothing came of McPeak's proposal.[87] It is unclear how serious McPeak was in his overture or whether he anticipated its rejection.

Commission on Roles and Missions Debates

The FY 1994 National Defense Authorization Act called for the Commission on Roles and Missions (CORM) to revisit and reevaluate the long-standing debates between the services about mission ownership.[88] The commission's specific role was to provide an analysis of the "efficacy and appropriateness for the post–Cold War era of current allocations among the Armed Forces of roles, missions, and functions."[89] Additionally, the commission was to examine alternative roles

---

world's air forces had neither SAMs nor AAA; 15 percent controlled SAMs; and 7 percent possessed both SAMs and AAA. That said, some highly capable air forces do have their own GBAD, most notably China and Russia, whose air forces have both SAMs and AAA.

[85] McPeak, 2017, pp. 313–315.

[86] Important to Sullivan's position, the Army aviation community viewed the Apache as a maneuver force for deep attack operations (which was the doctrine) during this time. For a discussion of the problems with this view, see David E. Johnson, *Learning Large Lessons: The Evolving Roles of Ground Power and Air Power in the Post-Cold War Era*, Santa Monica, Calif.: RAND Corporation, MG-405-1-AF, 2007, pp. 61–65.

[87] McPeak, 2017, pp. 313–315. We reviewed Sullivan's writings in an effort to see whether he offered additional perspectives. He did not pen a memoir, and we could find no reference to air base defense or related topics in any of the documentation that we reviewed.

[88] Pub. L. 103-160, National Defense Authorization Act for Fiscal Year 1994, November 30, 1993.

[89] Pub. L. 103-160, 1993.

and missions allocations and submit a series of recommendations for future functional distributions.[90]

The impetus for the commission's report, *Directions for Defense*,[91] came from Senator Sam Nunn, who was concerned with the high costs of duplication throughout the services—at one point describing a U.S. military with four air forces.[92] President Bill Clinton grew sympathetic to Nunn's protestations about redundancies, and, after a few failed attempts to address the issue, Congress eventually called for the CORM.[93] This was the first effort since Goldwater-Nichols to discuss roles and missions, and the final report offered insights on terminology and definitions and clarified some perceived areas of duplication.[94]

However, *Directions for Defense*, released in May 1995, was hardly a sweeping investigation of DoD's roles and missions.[95] The report offered narrow recommendations and lacked a conceptual framework to navigate the contentious issues on roles and missions. Key critiques came from senior policymakers and those on Capitol Hill. Many felt the recommendations did not address the primary issue the commission was created to examine: foundational roles and missions in DoD.[96] Although these critiques did not abate, the report clearly stated that it did not aim to outline a way ahead for roles and missions (although many had expected it would):

> What that means to those who read this report is that you are not going to see a listing of roles and missions disputes, or sharp Commission recommendations on how to resolve those disputes. You are not going to find a series of "put and take" statements that rearrange U.S. forces from one Service to the other. To have addressed our task in that way would have perpetuated the narrow institutional perspectives that inhibit development of a true joint warfighting perspective.[97]

In failing to address redundancies in roles and missions, the report was seen as a missed opportunity. Indeed, the report largely sidestepped the matter. While some privatization recommendations attempted to address the issue of overlap, the broader issues related to multiple

---

[90] Pub. L. 103-160, 1993.

[91] Commission on Roles and Missions of the Armed Forces, *Directions for Defense*, Washington, D.C.: U.S. Department of Defense, 1995.

[92] Raphael S. Cohen, *History and Politics of Defense Reviews*, Santa Monica, Calif.: RAND Corporation, RR-2278-AF, 2018, pp. 12–13.

[93] Eric Schmitt, "New Report on Long-Sought Goal: Efficiency in Military, *New York Times*, May 25, 1995.

[94] Although not central to our narrative here, the Goldwater-Nichols legislation, signed into law in 1986, made the most sweeping changes to DoD since the National Security Act of 1947 (Pub. L. 253, 1947). Goldwater-Nichols established a new military hierarchy that elevated the CJCS and the combatant commanders over the service chiefs. See Thornhill, 2019, pp. 175–178, for a more detailed discussion.

[95] By failing to fulfill the CORM's mission, the report continued to stoke service tensions (Bryan Bender, "USACOM Mission May Be Adjusted: CORM to Recommend New 'Functional' Command for Joint Force Training," *Inside the Army*, Vol. 7, No. 20, May 22, 1995).

[96] John T. Correll, "A New Look at Roles and Missions," *Air Force Magazine*, November 2008, p. 53.

[97] John White, "Preface," in Commission on Roles and Missions of the Armed Forces, 1995, p. 6.

medical, dental, and chaplain forces and to duplicative CAS capabilities were not resolved.[98] Three additional points of criticism evidence the disappointment of senior leadership following the report's release. The first comes from the report's lack of clear guidance on how to integrate or deconflict CAS across services. Second, the commission did not develop any general principles or a broad philosophical foundation for how to resolve disputes about roles and missions. Third, the report focused heavily on capabilities and how to properly equip unified commanders, to the detriment of examining the actual distribution of functions.

Directions for Defense did, however, make several contributions. It defined historically ambiguous terms (roles, missions, functions), proposed concrete steps to reduce inefficiencies, and dismissed red herrings often brought up in the roles and missions arena. The report defined key terms:

- *Roles* are the broad and enduring purposes specified by Congress in law for the Services and selected DoD components.
- *Missions* are the tasks assigned by the President or Secretary of Defense to the combatant commanders.
- *Functions* are specific responsibilities assigned by Congress, by the President, or by the Secretary of Defense to enable DOD components to fulfill the purposes for which they were established.
- *Capability* is the ability of a properly organized, trained, and equipped force to accomplish a particular mission or function.[99]

Even with such definitional guidance, inconsistencies persist. Most noticeably, the DoD dictionary does not mention *roles*, *functions*, or *capabilities*.[100] Its definition of *missions* remains consistent with the CORM report but does not indicate that a mission must apply to a combatant commander. Instead, the DoD dictionary defines *missions* as pertaining to lower-level units as well.[101] Despite discrepancies between the two, the definitions in each prove helpful in understanding the interplay of these concepts. The CORM report explained how these concepts interact, highlighting that they work in tandem rather than in isolation. The definitions provided further indicate the capabilities-focused approach the commission took in its report.

The report also noted that Senator Nunn's fear of a military with four air forces was misguided. The section of the executive summary labeled "'Problems' that are not Problems" suggested that most aircraft and air capabilities across the services represent complementary, rather than redundant, systems.[102] Finally, the report proposed ideas on how to maximize DoD's efforts on high-priority needs by cutting back on cost inefficiencies, specifically support and infrastructure.

---

[98] Cohen, 2018, p. 12.

[99] Quoted from Commission on Roles and Missions of the Armed Forces, 1995, p. 1-1.

[100] Joint Staff, *DOD Dictionary of Military and Associated Terms*, Washington, D.C., January 2020.

[101] Joint Staff, 2019, p. 148.

[102] Commission on Roles and Missions of the Armed Forces, 1995, p. ES-5.

The 1990s marked a return to militarywide discussions about the allocation of roles and missions. These debates came in a post–Goldwater-Nichols, post–Cold War, post–Gulf War environment that was ripe for the services to reevaluate their purposes and capabilities. This era also involved the military's shift to expeditionary offensive force structures and capabilities resulting from rear-area sanctuary. The inactivation of 32nd Army Air Defense Command in Europe and the emphasis on offensive operations were indicative of the reduced salience of air base defense. Not until roughly 2004 was there a return to the issue of air base defense and to the assignment of roles and missions that accompany it.

From the Global War on Terror to the Present

After September 11, 2001, DoD turned its collective energies toward combating terrorism. In the aftermath of the terrorist attacks, the U.S. invasion of Afghanistan in October 2001, and the subsequent invasion of Iraq in March 2003, the U.S. military turned its focus largely to conducting COIN operations and counterterrorism efforts. This focus on COIN and counterterrorism would persist essentially until the U.S. withdrawal of forces from Iraq in 2010. From an air base vulnerability point of view, these new wars in Central Asia and the Middle East would turn the clock back to the Vietnam era. Similar to the war in Southeast Asia, the U.S. Air Force would once again be forced to operate from bases within easy reach of enemy ground forces.

Simultaneous COIN operations in Iraq and Afghanistan tremendously stressed Army manning. In 2004, not long after the Iraq war began, Army and Air Force leadership terminated JSA 8, which had made the Army "responsible for providing forces for Air Base Ground Defense operations outside the boundaries of designated USAF bases and installations."[103] Importantly, the Army units assigned to this mission were to be under the operational control of the senior air commander. But in 2004, the Army simply lacked the ground forces to conduct both COIN and the defense of air bases. The abrogation of JSA 8 gave the Air Force the lead role for ground defense of air bases, both inside and outside the base perimeter.[104]

While the two services officially announced the end of JSA 8 because of operational requirements in Iraq and Afghanistan, the agreement had, in fact, been unofficially defunct for many years. The staff summary sheet put together to validate the end of the agreement noted that, after 1985, the JSA 8 working group met at least four times and agreed that the principles to

---

[103] Robert H. Holmes, "Validating the Abrogation of Joint Service Agreement 8," staff summary sheet, Washington, D.C.: Department of the Air Force, November 18, 2004; John P. Jumper and Peter J. Schoomaker, "Abrogation of Joint Service Agreement 8," memorandum of understanding, Washington, D.C.: Department of the Air Force and Department of the Army, March 24, 2005.

[104] Christensen, 2007, notes that GEN Peter Schoomaker, CSA, and Gen John Jumper, CSAF, both formally signed the memorandum officially abrogating both JSA 8 and JSA 9 in 2005. See Department of the Army and Department of the Air Force, *Abrogation of Joint Security Agreement 8*, Washington, D.C.: Department of Defense, March 24, 2005.

be set forth in JP 3-10 would provide doctrinal guidance. However, the agreement also stipulated it would remain in effect until rescinded or superseded by mutual service agreement and called for its review every two years. These reviews failed to transpire and were overtaken by changes in joint doctrine. In 1992, JP 3-10, *Doctrine for Rear Area Operations*, made the joint force commander–appointed installation commander responsible for the defense of the installation, regardless of service affiliation.[105] Thus, the agreement was never fully implemented or developed, essentially never coming to fruition. During the period of the agreement, the U.S. Army did not assign operational control of ground combat forces to an air commander.[106]

With the abrogation of JSA 8, the Air Force began to take on a larger role in defending its own air bases. In 2004, Lt Gen Walter E. Buchanan III, U.S. Central Command's coalition forces air component commander, received permission to organize, train, and equip the Air Force's first offensive ground combat task force, Task Force 1041. Its purpose, dubbed Operation Desert Safeside, was to conduct a 60-day operation to reduce standoff attacks against Logistics Support Area Anaconda—the Army's logistics base near Balad, Iraq, and also home to the Air Force's busiest airfield.[107] In 2004 alone, indirect fire attacks on Joint Base Balad occurred nearly once a day. In recounting the situation at Balad prior to Operation Desert Safeside, Buchanan noted that

> [i]t was not that the Army wasn't concerned about the insurgent attacks on Balad; it was. However, Balad AB was only one of many important sites in a huge AO owned by the First Infantry Division, Second Brigade Combat Team [BCT]. The 2 BCT AO [area of operation] stretched from Tikrit to Baghdad, and it was taking casualties regularly. It did not have the manpower to focus operations specifically on defeating the standoff threat around any base or site including Balad AB.[108]

Operation Desert Safeside was the first time since the war in Vietnam that the Air Force Security Force was given the mission of fighting outside the wire in defense of an air base. It was also one of only two times that the Air Force has been given an offensive ground combat mission.[109] The operation lasted 120 days and was generally deemed a success. It eliminated 17 high-value targets, 98 insurgents, and eight major weapon caches. The task force also reduced the number of standoff attacks against Joint Base Balad to "nearly zero."[110]

Internalizing and evaluating the early lessons in Iraq on air base defense, the Air Force would refine its tactics, techniques, and procedures. This resulted in the concept of *integrated*

---

[105] JP 3-10, *Doctrine for Rear Area Operations*, Washington, D.C.: Joint Staff, 1992; Holmes, 2004.

[106] Christensen, 2007. It is likely that the Army did not have any cause to do so either.

[107] Walter Buchannan, III, Bradley D. Spacy, Chris Bargery, Glen E. Christensen, and Armand "Beaux" Lyons, "Outside the Wire: Recollections of Operations Desert Safeside and TF 1041," in Caudill, 2019.

[108] Buchannan et al., 2019, p. 147.

[109] Buchannan et al., 2019, p. 140.

[110] Christensen, 2007, p. 7.

defense.[111] Joint Base Balad would once again provide the setting for the Air Force to test its new concept for defending its bases. In 2006, insurgents had mounted over 400 attacks on the base. In 2008, the Air Force was designated as base operating support integrator for Joint Base Balad. In this capacity, the Air Force was responsible for defending the base and its assigned forces, including the conduct of counter–indirect fire operations outside the base perimeter.[112] Prior to this time, the base defense strategy was largely reactive, consisting of chasing indirect fire shooters or directing counterbattery fire at the incoming fire's point of origin. Success was mixed.

When the Air Force assumed the base operating support integrator role in 2008, it adopted a more proactive strategy that consisted of the largest combat deployment of security forces since Vietnam. It also committed ground intelligence analysts combined with residual air assets in an effort to map the human terrain outside the base perimeter. The proactive patrolling of security forces freed local forces to conduct enhanced COIN operations. These combined results were largely fruitful: The number of insurgent attacks fell by 75 percent and proved much less effective. Successes notwithstanding, ambiguity persisted. One analysis of the operation noted a lack of clarity concerning who was responsible for protection beyond the base perimeter—an ambiguity that persisted at the "most senior levels of USAF leadership."[113] The confusion was due in no small part to roles and missions disagreements about what constituted defensive operations in the joint community.[114] The Air Force also had not fully embraced or resourced integrated defenses, especially the training and equipment requirements.

The Air Force would again implement integrated defense—this time in Afghanistan's Helmand Province, at NATO's busiest base. In spring 2010, insurgents conducted a complex attack on Bagram AB, killing a U.S. contractor and wounding nine coalition members.

Wearing U.S. uniforms, the attackers began their assault with indirect fires and then attempted to breach the base perimeter.[115] Discussions between the U.S. Army and Air Force over the base's security began later that year. Progress on the base's defense proceeded in phases. By the following July, the 455th Air Expeditionary Wing stepped up perimeter defenses and focused on installation entry control points. At the end of 2010, the Air Force negotiated limited maneuver operations up to 500 m outside the base perimeter, an effort initially undertaken to fill the void created by a reduction in the Army presence in the area. In 2011, a

---

[111] The Air Force originally published AFTTP 3-10.1, *Integrated Base Defense*, in 2004 and then AFTTP 31-1, *Integrated Defense*, in 2007. See Milner, 2014, pp. 217–218. The concept calls for a combination of active and passive defense measures, employed across the ground dimension of the operational environment. It is a capabilities-based construct that emphasizes intelligence collection in the operating environment.

[112] Milner, 2014, p. 218.

[113] Milner, 2014, pp. 218–219.

[114] Milner, 2014, pp. 218–219.

[115] Briar, 2014, p. 269.

lack of manpower forced the Army to discontinue dedicated exterior patrols of Bagram. From July 2011 to February 2012, the Air Force again expanded its role by moving further from the base perimeter. In May 2012, Task Force 1/455 was formally activated to integrate all force-protection efforts under a single brigade-level commander, an Air Force Security Force colonel. At the time the task force was deactivated in 2013, it was responsible for the largest outside-the-wire combat mission in Air Force history.[116] The initiative successfully reduced the number of indirect fire attacks and completely eliminated insider attacks during a period of sometimes intense enemy activity elsewhere in the region.[117]

Looking Toward the Future

While the Air Force experience in Iraq and Afghanistan during the past two decades demonstrates its ability to defend its bases in a COIN setting, new challenges related to peer competition are already changing the debate on air base defense. As discussed in Chapter 2, the United States has significant shortfalls in AMD that put fixed facilities, such as air bases, at risk.

The Missile Defense Agency (MDA) has taken the lead on ballistic missile defense; the services have played secondary roles. MDA is responsible for developing, testing, and fielding an integrated, layered ballistic missile defense system in defense of the United States, its deployed forces, and its partners and allies. In spite of some successes, MDA has made mixed progress in achieving its delivery and testing goals. DoD oversight remains a problem. A 2008 MDA study in response to congressional direction noted a major issue was the process and timing of transferring responsibility for operations, maintenance, and follow-on procurement from MDA to the respective services.[118] A more recent report from the U.S. Government Accountability Office (GAO) echoes similar concerns. It states that, while MDA continued to deliver assets to the military services, system-level integrated capabilities were delayed and delivered with performance limitations. Additionally, the GAO described challenges in MDA's processes for communicating the extent and limitations of integrated capabilities when they are delivered. Consequently, the military lacked full insight on the capabilities MDA delivers.[119]

---

[116] It was also the first time the Air Force had been the responsible battlespace owner at the brigade level within a combat zone. Our account draws from Eric K. Rundquist and Raymond J. Fortner, "An Airman Reports: Task Force 455 and the Defense of Bagram Airfield, Afghanistan," in Caudill, 2019.

[117] Rundquist and Fortner, 2019, pp. 133–134.

[118] A lead service is designated for each ballistic missile defense system component except C2, battle management, and communications. See Larry D. Welch, D. L. Briggs, R. D. Bleach, G. H. Canavan, S. L. Clark-Sestak, R. W. Constantine, C. W. Cook, M. S. Fries, D. E. Frost, D. R. Graham, D. J. Keane, S. D. Kramer, P. L. Major, C. A. Primmerman, J. M. Ruddy, G. R. Schneiter, J. M. Seng, R. M. Stein, S. D. Weiner, and J. D. Williams, *Study on the Mission, Roles, and Structure of the Missile Defense Agency (MDA)*, Alexandria, Va.: Institute for Defense Analyses, IDA Paper P-4374, 2008.

[119] GAO, *Missile Defense: Some Progress Delivering Capabilities, but Challenges with Testing Transparency and Requirements Development Need to Be Addressed*, Washington, D.C.: GAO-15-381, May 2017.

Regarding cruise missile defenses, MDA's role has been minimal, although, in 2018 testimony before the Senate, its director included hypersonic and cruise missiles among the threats that MDA was prepared to address.[120] It certainly would be good to have MDA's resources contributing to cruise missile defense, and there may be opportunities for the Air Force to partner productively with MDA in advancing key technologies. Nevertheless, Air Force leaders should proceed cautiously for several reasons.

First, as noted in Chapter 3, cruise missile defense is not primarily a technology problem but one of institutional priorities and proper funding. Second, the missile defense community has shown little appreciation of or interest in the cruise missile defense problem. For example, the 2019 Missile Defense Review largely ignored the problem: "The only cruise missile defense plan described by the MDR is a preexisting, three-phase effort for the U.S. national capital region. This initiative is important, but cruise missiles threaten U.S. forward-deployed forces around the world, too."[121] Third, MDA has proposed "a new multibillion-dollar project to develop and field a defense against hypersonic weapons," a technical challenge more in keeping with MDA's history, culture, and expertise.[122] This project took a step forward on December 5, 2019, when MDA "announced in a public notice plans for a Hypersonic Defense Regional Glide Phase Weapon System Prototype Project, an effort that will be executed using other transaction authorities to rapidly design, develop and demonstrate an initial system."[123]

In announcing the proposal, an MDA assistant director claimed that the United States had solved the cruise missile defense problem—which suggests a certain institutional disinterest.[124] More telling yet, in a recent Defense News interview, the new director of MDA did not include air or cruise missile defense in his list of priorities and did not mention cruise missile defense at any time during the interview.[125] Furthermore, a 2008 IDA study on MDA's roles and missions noted that "there is little convergence between the functional demands of ballistic missile defense and cruise missile defense and only limited opportunities for dual use among host systems" and recommended against assigning MDA the responsibility for cruise missile defense.[126] In sum, MDA does not appear to be a necessary or natural partner to advance the capabilities needed for point AMD of air bases.

---

[120] Rehberg and Gunzinger, 2018, p. 10.

[121] Thomas Karako, "The Missile Defense Review: Insufficient for Complex and Integrated Attack," *Strategic Studies Quarterly*, Summer 2019, p. 11.

[122] Jason Sherman, "DOD Readies New Weapon System Plan for 'Regional' Hypersonic Defense," Inside Defense website, December 11, 2019d.

[123] Sherman, 2019d.

[124] Jason Sherman, "MDA Forwards Proposal for New Hypersonic Defense Program to Pentagon for Review," Inside Defense website, August 7, 2019c.

[125] Jen Judson, "U.S. Missile Defense Agency Boss Reveals His Goals, Challenges on the Job," Inside Defense website, August 19, 2019c.

[126] Welch et al., 2008, p. VIII-1.

Key Insights

This brief survey demonstrated that the specific function of defending U.S. air bases has a circuitous history, with several problems left unresolved as of this writing. We highlight some important issues that emerge from this review.

Persistently an Air Force concern, the defense of air bases has not enjoyed similar prioritization by other services, particularly the Army. This remains an unfortunate fact for the Air Force. Since World War II, the Army has been responsible for air base point defense, but its performance of the task has consistently failed to fully satisfy Air Force needs and concerns. Our historical review suggests several reasons for this. Resource limitations and budget constraints have, time and time again, forced the services to prioritize and make concomitant trade-offs.

The Army—America's ground force—has consistently and not unexpectedly prioritized other tasks and other functions over air base defense. This was particularly so during the Vietnam War, when air bases continually came under attack, and in the 1980s, when the Army prioritized area defenses (i.e., Patriot) over point defense capabilities. The resurgent Russian threat in Europe has only reinforced the Army's tendency to prioritize protection of maneuver forces. Although Congress directed the Army to acquire the Israeli Iron Dome system to provide a near-term capability against cruise missile threats to rear facilities, BG Brian Gibson, Director of the Air and Missile Defense Cross Functional Team, has made it clear that the Army has other plans. Speaking to reporters on October 18, 2019, Gibson observed that Iron Dome "was developed for a very specific threat and it does incredible things," but "we intend to operate it differently—we intend to operate it in support of an Army on the move. It's not just going to be static."[127]

The ebb and flow of conflict throughout the decades has also affected this relationship. Interwar service initiatives to address air base defense shortcomings have proved either short-lived for lack of sustainable interest or out of reach because of diverse and conflicting perspectives. Priorities for air base defense have diverged the most during U.S. wars and in active battlespaces, and the Air Force has assumed more responsibility for the function. However, wartime adaptations were either (1) temporary arrangements in which the Air Force assumed increased (ground) responsibilities—as was the case during operations Iraqi Freedom and Enduring Freedom—or (2) muddled and haphazard—as was the case during Vietnam.

Another important factor to emerge from our review of air base defense is its lack of prominence at the highest levels of discussions about roles and missions. In the 1950s, key Army and Air Force debates on roles and missions centered on missiles and air defense.

Unfortunately, the absence continues today. The most recent iteration of DoD Directive 5100.01, the current document outlining service functions, offers little more clarity on the task

---

[127] Paul McLeary, "U.S. Army Signals Israel's Iron Dome Isn't the Answer," Breaking Defense website, October 15, 2019b.

than did the Key West document.[128] In sum, there has been remarkably scant progress in high-level DoD guidance toward defining and clarifying service responsibilities for defending air bases. That this functional gap has persisted for so many decades—through several wars, various DoD reforms, and numerous interservice initiatives—suggests that, without institutional impetus for change at the highest levels, it may well endure much longer.

In Chapter 6, we turn to organizational strategies that the Air Force might pursue to overcome the institutional constraints identified here.

---

[128] DoD Directive 5100.01, *Functions of the Department of Defense and Its Major Components*, 2010.

6. Organizational Strategies to Improve U.S. Air Force Air Base Defense Capabilities

Prior chapters explored the evolving threat to air bases, concepts and technology options to enhance air base defenses, and the evolution of R&F assignments related to air base defense. In this chapter, we argue that, although there are many contributing factors, the institutional root cause of air base defense shortfalls is the misalignment of air base defense responsibilities and organizational stakes. We then assess seven potential U.S. Air Force COAs with respect to their ability to correct this misalignment and other benefits that might flow from them.

A Framework to Better Align Service Responsibilities for Air Base Defense

As we detailed in the previous two chapters, the Air Force and Army have disagreed over their respective responsibilities for air base defense for most of the past 70 years. In the 1950s, they disputed the utility of AAA for air defense of SAC bases and disagreed over which SAM system was best for point air defense. In the 1960s and early 1970s, the Air Force, suffering from ongoing and costly rocket attacks on its air bases in Vietnam, repeatedly and unsuccessfully asked the theater commander in Vietnam, as well the CSA, to dedicate infantry forces to the ground defense of Air Force bases. The 1980s were the one period when the services did work cooperatively to establish and clarify their respective roles for base defense. The two service chiefs signed a 1984 MOU on point air defense and JSA 8, delineating respective service responsibilities for ground defense of bases.

Air base defense lost salience for both services during the early post–Cold War years, when the United States enjoyed unchallenged military superiority. During the 1990s, the MOU on air defense (which required periodic renewal) expired unnoticed. In the following decade, force structure shortfalls during Operation Iraqi Freedom led the Army to abrogate JSA 8 in 2004, forcing the Air Force to provide for both perimeter and external defense of air bases; similar arrangements were established for the defense of Bagram AB in Afghanistan.

Today, the U.S. Army is responsible for providing point AMD for Air Force bases and other fixed facilities, but years of neglect from both services have resulted in capability and capacity shortfalls, especially against cruise missiles. Army leadership has understandably prioritized mobile short-range air defense for its maneuver units over fixed facility defenses (e.g., IFPC). Mobile short-range air defense is among the Army's top AMD priorities, because it has little capability. However, the Army has also recently emphasized the importance of other AMD systems, including IFPC, to focus on higher-end threats, such as cruise missiles. It is also complying with Congress' mandate that it field Iron Dome to address the rocket and cruise missile threat. But the Army is largely ambivalent about Iron Dome, and competing demands

97

will likely force hard procurement decisions.[1] Perhaps of greater concern than these difficult resource allocation choices is that air base defense is not even among the issues that the Army AMD community considers worthy of including in its 2028 vision document, which does not refer to *air base*, *airfield*, *USAF*, *Air Force*, or *air base defense* anywhere.[2] By contrast, although the Air Force increasingly highlights the threat enemy missile systems pose, it cannot compel the Army to provide these forces.[3]

We emphasize here a view of R&F that places (at least in part) the root cause of these disputes on a misalignment of responsibilities and organizational interests. This can be seen most clearly in the framework illustrated in Table 6.1. For a given function, the framework asks four questions: (1) Which service is assigned the task? (2) Which service has the greatest stake in the outcome? (3) Is the activity a service priority? (4) Does the responsible service have forces dedicated to the task? Ideally, the service assigned the responsibility would also perceive substantial stakes and, consequently, would make accomplishing the task a service priority. Such prioritization concerns would not matter if the joint requirements process ensured that all important requirements were met. Unfortunately, the process does not do this. The military services have always taken the lead in advocating for and resourcing programs in their respective mission areas. The result is that service priorities tend to align with the stakes each organization has in fulfilling a particular function.

Table 6.1 offers two contrasting examples, one in which service responsibilities are well aligned with their priorities and one in which they are not. In Example 1, when the fleet is at sea, defense against air and missile attack is entirely a Navy responsibility, supplemented by Marine Corps aviation when it is embarked. This is appropriate, because the Navy has the greatest stake in mission success. The task is also a service priority, resulting in significant investments in airborne early warning, battle networks, interceptor aircraft, and advanced air and ballistic missile defense systems. The Navy also dedicates forces (e.g., the E-2 Hawkeye early warning aircraft) to fleet AMD. While Marine Corps aviation shares in the mission when it is afloat, the Navy develops, fields, and advocates its own AMD programs.

In contrast, service responsibilities for ground-based air defense of Air Force bases are misaligned (Example 2). The Army is assigned the task but does not perceive high stakes and, consequently, has made it a low priority with no forces dedicated to air base point air defense.

---

[1] For a related discussion, see Sydney J. Freedberg, Jr., "Army Reboots Cruise Missile Defense: IFPC & Iron Dome," Breaking Defense website, March 11, 2019a.

[2] Based on a keyword search of a PDF of the vision document. See U.S. Army Space and Missile Defense Command, *Army Air and Missile Defense 2028*, Huntsville, Ala.: USASMDC/ARSTRAT, 2019. To be fair, the document includes a direct reference to the protection of "critical fixed and semi-fixed assets," which would include air bases among other assets (p. 11).

[3] It is important to acknowledge that, ultimately, the theater commander determines force allocation, and the services are the force providers that satisfy these demands. The Joint Staff and the Joint Capabilities Integration and Development System (JCIDS) decide how to close capability gaps. We highlight this in the following section.

**Table 6.1. Examples of Well Aligned and Misaligned
Service Responsibilities for Air Defense**

	Example 1: Fleet Air Defense Afloat		Example 2: Ground-Based Air Defense of Air Force Bases	
	Navy	Marines	Army	Air Force
Service assigned responsibility?	Yes	Shared with Navy when afloat	Yes	No
Service with greatest stakes?	Yes	Shared with Navy when afloat	No	Yes
Service priority?	Yes	No	No	Growing
Dedicated force structure?	Yes	When afloat	No	No
	Well aligned		Not well aligned	

The Air Force perceives great stakes in the task, and air base defense is slowly growing in importance to airmen, but R&F decisions dating back to 1956 prohibit the Air Force from developing and deploying short-range ground-based SAMs. The only Air Force force structure dedicated to base defense is in its security forces, whose focus is on ground threats. The Air Force Security Force is testing some prototype counter-UAS systems but has no other mandate or capability for air defense.

This approach is not without flaws. If the stakes and priorities argument is as strong as we suggest, it should be applied elsewhere. For example, the Army might argue that it has the highest stakes in CAS or tactical airlift and that, in its view, these functions have never been high priorities for the U.S. Air Force. Following this logic, the Army might argue that it should have the freedom to develop its own CAS and tactical airlift capabilities. This actually is not that far off from actual debates over the acquisition of Army rotary-wing attack and lift aircraft and periodic Army efforts to acquire tactical fixed-wing airlift (e.g., the C-7 Caribou). Applying this approach broadly could result in significant losses in efficiency as functional capabilities are spread piecemeal across the services. For example, it is much more efficient for the Air Force to provide C-130 tactical airlifters to meet ground-force lift needs than for the Army to acquire and sustain a small C-130 fleet of its own. As the joint force moves toward joint all-domain operations, the services will have to become even more interdependent. In a joint all-domain operations world, it should matter less who owns what function.

We also recognize that giving the Air Force the ground-based air defense mission would not guarantee that Air Force leaders would make air base defense a priority over competing demands for resources or that they would dedicate forces for AMD. The centrality of the offensive in airpower thought has produced an enduring institutional preference for investments in offensive, rather than defensive, capabilities. Whether current and projected threats are sufficient to overcome this reluctance to make major investments in defenses remains to be seen. Assigning air base defense solely to the Air Force would simply give airmen the authority that they have long sought, a necessary first step in improving air base defense capabilities. But it would also present a difficult question to the Air Force: How much structure and capability would it be willing to buy, if it owned the function, at the expense of other programs and priorities?

In sum, we propose that service stakes and priorities should be considered alongside efficiency considerations during discussions of R&F. Roles, missions, and functional assignments should not be barriers to innovation or to correcting critical warfighting shortfalls. Although efficiency matters, it is less important than giving services the freedom to pursue essential capabilities and innovations.

Alternative Air Force Courses of Action

In this section, we consider seven potential COAs that Air Force leaders might pursue to improve air base defenses. Although some of the COAs are mutually exclusive, several of them could be pursued in various combinations.

COA 1: Pursue Air Base Defense Options That Are Clearly Within Air Force Purview

The most straightforward option available to Air Force leaders is to pursue options that are already in line with the Air Force's assigned functions. It already is doing so to some degree, updating civil engineer runway repair capabilities to better align with projected threats, developing various concepts for deployable protective aircraft shelters, designing a hardened aircraft shelter for large aircraft, and practicing dispersed operating concepts in both Europe and Asia. But the scale and pace of these programs is modest, well below the effort made the last time the Air Force faced a serious threat to air bases.

During the Cold War, the Air Force embarked on a massive effort to harden air bases in Europe and Asia. Recognizing the risk to unsheltered aircraft, the Air Force built close to 1,000 hardened aircraft shelters for fighters at forward bases. The shelters were fully enclosed, providing excellent protection from all but a direct hit from a unitary weapon. In Europe, larger aircraft, such as tankers, were based in Spain and the UK, outside the range of most threat systems. Various concepts for distributed operations were also pursued, although not to the degree that hardening was embraced. The Air Force also invested in improved runway repair capabilities and hardened infrastructure, such as command posts and fuel and munition storage.[4]

Although the particular infrastructure investments will differ in 2020, a serious and sustained Air Force effort to improve air base resiliency would include increasing the number of personnel (and units) dedicated to air base resiliency (e.g., security forces, civil engineers), buying large numbers of fuel bladders and other support equipment necessary for distributed operations, expanding the number of rapid runway repair teams, buying and forward deploying

---

[4] For more on these Cold War programs, see Christopher J. Bowie, "The Lessons of Salty Demo," *Air Force Magazine*, March 2009, pp. 54–58; Benson, 1981; Weitz, 2001; and Vick, 2015a.

expeditionary style protective shelters, and hardening MOBs where appropriate (e.g., building underground command posts, improving the blast resistance of aboveground structures, and renovating existing hardened aircraft shelters and/or building new ones).

COA 2: Use Existing Joint Processes to Address Cruise Missile Defense Shortfalls

The 1986 Goldwater-Nichols Act and subsequent DoD organizational reforms created joint processes to ensure that warfighters' needs are met through the requirements, research, development, acquisition, and sustainment processes. Specifically, JCIDS; the Planning, Programming, Budgeting, and Execution System (PPBE); and the Defense Acquisition System (DAS) provide venues to identify capability shortfalls, propose solutions, develop new technologies, and acquire new capabilities as needed.[5]

In theory, the PACAF, USAFE, or any other air component commander can articulate the requirement for improved air base air defenses to their combatant commander who, in turn, can generate a requirement that, ultimately, would be given to the Army, as the ground-based AMD provider, to fulfill. The actual process is vastly more complicated, involving dozens of potential failure points, most notably Joint Requirements Oversight Council and OSD approval; Army leadership agreement to expand its SHORAD force structure (and assign it to air bases); and, finally, congressional funding of the new or additional weapon systems, air defense battalions, and manpower billets. These processes are complex and typically slow, but they exist to address capability and capacity shortfalls.

We have no independent means of determining how often or aggressively PACAF, USAFE, or other air component commanders have sought to correct ground-based air defense shortfalls through these mechanisms. The only evidence we have is the lack of validated joint requirements or funded programs to improve air base AMD. The PACAF commander raised air base vulnerability as a joint problem in 2008, and there have since been many joint war games and analyses coming to similar conclusions. But so far, there have been no capability improvements. Indeed, as noted earlier, Congress was so frustrated by lack of progress on the IFPC program that it directed OSD and the Army to field some cruise missile defense capability for fixed facilities by 2020.

If USAFE and PACAF leaders have not yet clearly expressed this requirement to the U.S. European Command and PACOM commanders, they certainly should.

---

[5] For more on JCIDS, PPBE, and the DAS, see John Rausch, "Joint Capability Integration & Development System Overview: New Manual and Sustainment Key Performance Parameters," briefing slides, Washington, D.C.: Joint Staff, August 31, 2018, and CJCS Instruction 5123.01H, *Charter of the Joint Requirements Oversight Council (JROC) and Implementation of the Joint Capabilities Integration and Development System (JCIDS)*, Washington, D.C.: Joint Staff, August 31, 2018.

Thanks to reviewer David Johnson for recommending inclusion of this COA.

We are, however, skeptical that the JCIDS, PPBE, and DAS processes will produce improved ground-based air defense unless there also is strong and sustained support from the Army leadership. Colin Jones and Alexander Kirss captured the essence of the problem:

> The root of the . . . problem—JCIDS' inability to effect change below the joint level—is that the Joint Staff is expressly barred by the Goldwater-Nichols Act and current U.S. Code in Title 10 from unilaterally making certain major changes to how the military operates. As a Joint Staff–owned process, JCIDS must rely on the services to implement its recommendations. As a result, the joint force is actually not built as a joint force—it is a product of the agglomeration of service-parochial capabilities. To be sure, the joint force has become more joint over the past few years. Still, the divide between the services and joint equities—for instance, bureaucratic fights over unique capability needs or who "owns" what missions—hobble the department's ability to build and fight as one force. JCIDS is intended to iron out these issues, but without the authority to directly enforce change, it is less effective than it could be.[6]

In summary, one strength of this COA (if successful) is that it could address air base defense shortfalls without requiring the Air Force to create new forces and manpower billets, develop training programs, and procure new weapon systems. Although Air Force leaders are concerned about air base vulnerability, it is not clear that the institution is ready to shift resources to acquire such capabilities.[7] On the other hand, the JCIDS process cannot compel the Army to meet Air Force air base defense needs. Thus, the process depends on the Army leadership making it a priority. As noted elsewhere in the report, although the Army does prioritize AMD more broadly, it has not historically emphasized air base air defense.[8]

---

[6] Colin Jones and Alexander Kirss, "Some Modest Proposals for Defense Department Requirements Reform," War on the Rocks website, August 23, 2018. For a related critique of JCIDS, see Jarrett Lane and Michelle Johnson, "Failures of Imagination: The Military's Biggest Acquisition Challenge," War on the Rocks website, April 3, 2018.

[7] Although airmen around the world have recognized the importance of attacking enemy airfields and defending their own since World War I, this has been conceptualized in airpower theory as primarily an air battle. Base-level defenses (e.g., against ground or air threats) are rarely discussed in airpower writings. For example, a word search in five recent major Air Force strategic documents—*A Vision for the United States Air Force* (2013), *Global Vigilance, Global Reach, Global Power for America* (2013), *Global Horizons* (2013), *America's Air Force a Call to the Future* (2014), *Air Force Future Operating Concept* (2015)—found not a single mention of either "air base" or "air base defense." Additionally, a review of *Air University Review*, *Air Power Journal*, and *Air and Space Power Journal* issues published between 1947 and 2019 found relatively few references to local base defense. For example, only eight of the 59 issues published between 2010 and 2019 contained any reference to local base defense. See U.S. Air Force, *Global Vigilance, Global Reach, Global Power for America*, Washington, D.C.: Department of the Air Force 2013b; U.S. Air Force, *The World's Greatest Air Force, Powered by Airmen, Fueled by Innovation: A Vision for the United States Air Force*, Washington, D.C.: Department of the Air Force, 2013c; U.S. Air Force, *Global Horizons Final Report United States Air Force Global Science and Technology Vision*, Washington, D.C.: Department of the Air Force, June 21, 2013d; U.S. Air Force, *America's Air Force: A Call to the Future*, Washington, D.C.: Department of the Air Force, July 2014; U.S. Air Force, *Air Force Future Operating Concept: A View of the Air Force in 2035*, Washington, D.C.: Department of the Air Force, September 2015.

[8] A review of the *Army Air Defense Journal* found that only six of 150 issues published between 1948 and 2006 mentioned air base defense.

COA 3: Bypass Roles and Functions Roadblocks Through Innovative Use of Technology

Today, in 2020, there is a great openness to innovation, partnering, and research and development (R&D) risk-taking in the U.S. defense enterprise. Whether such initiatives as the Strategic Capabilities Office, Defense Innovation Board, and the Air Force's innovation hub (AFWERX) ultimately achieve their desired outcomes, the pace of experiments and the extent of partnering are impressive.[9] New experiments and successful prototyping of HPM, laser, and other advanced technologies are being announced every week. Examples include the Air Force's THOR HPM counterdrone system and the multiple tactical lasers that the Navy, Marine Corps, and the Army are developing or have fielded as prototypes.

Counterdrone concepts and systems appear to fall into a class of capabilities in which the services are willing to partner to advance the state of the art and rapidly field interim capabilities tailored to their particular needs. It helps that there is a sense of urgency. Thus, the Army has not objected to Air Force efforts to develop counterdrone defenses for air bases, even though the mission nominally belongs to the Army as the owner of SHORAD systems. Laser and HPM counterdrone systems are of particular interest because the urgent need for this capability is giving momentum to directed-energy R&D, specifically to scaling up these systems so that they have real potential against cruise missiles.

There is no guarantee how this will play out from the perspective of R&F. However, if the Air Force is either the leader or a major partner in advancing these cruise missile defense technologies, it may have the opportunity to field prototype systems without triggering R&F antibodies. Controversies about R&F often center on weapons or technology. For example, an Air Force proposal to field its own ground-based SAMs could get caught up in R&F debates within the Pentagon or on Capitol Hill.[10] In contrast, innovative approaches and technologies, particularly in experiments and prototype systems, appear to bypass such roadblocks in today's environment. That does not necessarily mean that the Air Force would be given the green light to field dozens of laser weapon systems for cruise missile defense, but the current enthusiasm for innovation suggests that is much more likely than if the Air Force sought to procure a more conventional system. Furthermore, deployment of prototype systems has the potential to create facts on the ground, changing defense community perceptions about what is acceptable or normal simply by taking the initiative to develop new capabilities, however nascent or limited.

---

[9] For more on these innovation initiatives, see U.S. Department of Defense, Defense Innovation Board website, undated; Cheryl Pellerin, "DoD Strategic Capabilities Office Is Near-Term Part of Third Offset," U.S. Department of Defense website, November 3, 2016; and U.S. Air Force, AFWERX website, undated.

[10] An alternative possibility is that the Army would be willing to let the Air Force provide for air base defense, not unlike what happened in Afghanistan and Iraq. This would obviously depend on the context and whether the country were at war.

Another possibility is that all these creative partnerships among the services, the Defense Advanced Research Projects Agency, and the private sector are breaking down antiquated and artificial R&F barriers, offering each service more room to innovate where it sees an opportunity to contribute to a critical joint mission or to meet a service-specific need. This suggests that the Air Force should enthusiastically embrace all partnering opportunities where it sees a match between its own R&D efforts and an operational need. As the Air Force 2013 Vision observed: "The story of the Air Force is a story of innovation."[11] The Air Force, the service most strongly associated with advanced technologies in at least one public opinion survey, has little to fear from breaking down such barriers and much to gain from a greater emphasis on advanced technologies and innovation, particularly in gaining more autonomy to push air base defense innovations.[12]

COA 4: Propose a New MOU with the Army to Clarify That Ground-Based Air Defense of Air Bases Is an Army Responsibility

This is the first of three potential COAs built around a new Army–Air Force MOU. As discussed in Chapters 4 and 5, there are two precedents for an MOU specifying that some aspect of air base defense is an Army responsibility.

The first instance was the Air Force proposal in 1971 for an MOU that made explicit that the Army had responsibility for protecting Air Force bases from ground attack, whether the attacker used direct fire or indirect fire (standoff) weapons. This proposal resulted from Air Force frustration with U.S. Army reluctance to provide ground forces for air base defense during the (then still ongoing) Vietnam War. Army leadership rejected the overture, responding that the joint doctrine of the time was sufficiently clear on Army responsibilities and that no further clarification was needed.

The second precedent was JSA 8, signed in 1985. This specified that the Air Force was responsible for internal and perimeter security of Air Force bases; the Army was responsible for the battlespace surrounding the air base. This JSA was formally abrogated in 2004 when the Army found that manpower shortfalls prevented it from providing perimeter defense at Joint Base Balad in Iraq.

Neither of these precedents offers much hope for such agreements. Even if the Army were to sign one along these lines, it is unclear how that would differ from the status quo where it has been long understood within DoD that the Army has the responsibility for air base SHORAD

---

[11] U.S. Air Force, 2013c, p. 4.

[12] In a public opinion survey that the RAND American Life Panel conducted in May 2014, 50 percent of respondents identified the Air Force as the service they most associate with advanced technologies. "No opinion" was second at 21 percent, and the U.S. Navy was third at 17 percent. See Alan J. Vick, *Proclaiming Airpower: Air Force Narratives and American Public Opinion from 1917 to 2014*, Santa Monica, Calif.: RAND Corporation, RR-1044-AF, 2015b, p. 129.

systems, yet competing priorities and resource limitations have resulted in a growing gap between the threats and forces available for point defense.

COA 5: Propose a New MOU with the Army to Establish Ground-Based Air Defense as a Joint Air Force–Army Responsibility

An MOU that established ground-based air defense as a joint Air Force–Army responsibility is potentially a more promising and workable approach. The 1984 MOU on air base air defense (discussed in Chapter 5) offers a precedent. It directed that the Air Force would participate in the force planning process, providing its requirements to the Army. It also stated that the Air Force would fully support Army programming efforts to build and sustain the needed air defense forces, making the two services partners in advocating for these capabilities in internal DoD deliberations, congressional testimony, and public statements. Finally, the MOU stipulated that, if the Army were unable to meet requirements, the Air Force could pursue other partnering options or, if necessary, develop its own organic point air defense capabilities. Unfortunately, this MOU became active in the midst of the twin Army SHORAD debacles. The Roland missile program had been severely curtailed in 1981, deploying only one Roland-equipped National Guard battalion in 1985, only to disband the unit in 1988. The Sergeant York gun was canceled in 1985, a year after the MOU was signed. Given these developments, one might have expected that somewhere between the Sergeant York cancellation in 1985 and the Berlin Wall falling in 1989, the Air Force would have acted on the MOU clause giving it authority to pursue its own organic point defense capability. We were unable to find any historical records of Air Force actions along these lines in that period. Had the Cold War continued, perhaps the Air Force would have pursued this option.

In the event, neither the Army nor the Air Force was particularly interested in SHORAD systems once the Cold War ended, and it is not likely that OSD or Congress would have supported major funding for new air defense weapons in the 1990s, a time of military drawdown. The MOU was terminated automatically by the services' failure to renew it, possibly as early as 1988 but certainly no later than the early 1990s.

Would a new version of the MOU help solve the Air Force's point air defense problem? Perhaps, but despite the concrete and useful steps dictated by the MOU (e.g., involving the Air Force in the requirements process) and the remarkable clause allowing the Air Force to pursue its own SHORAD capabilities if needed, the MOU resulted in no measurable improvements in point air defense at air bases. It is not clear that a new version would necessarily put any more pressure on the Army to deliver than the 1984 MOU did. An MOU with an exit clause is worth pursuing only if the Air Force intends to act on it. In that case, it might be more efficient for the Air Force to seek that authority directly rather than buried in another joint MOU. A new MOU establishing joint responsibilities for this mission also fails to get at the root of the problem: the asymmetry in stakes discussed earlier in this chapter. Given the leadership attention and staff effort necessary to successfully negotiate an interservice MOU, this option seems to offer relatively little payoff for the investment.

COA 6: Propose a New MOU with the Army to Establish Ground-Based Air Defense of Air Bases as an Air Force Responsibility

Of the three MOU options discussed here, this is the only one that fully addresses the central problem of misaligned service responsibilities and stakes. An MOU along these lines might state that resource limitations prevent the Army from meeting these Air Force requirements and that the Air Force is therefore authorized to pursue its own point air defense capabilities. This is essentially the logic the Army used in abrogating JSA 8 for ground defense of air bases—that the Army did not have the resources to meet the commitment. Given past Air Force attempts to own most Army air defense systems, the MOU would need to define limits on what capabilities the Air Force would be authorized to pursue.[13] For example, the MOU might limit the range and capabilities of the systems or specify that the number of systems acquired would have to be tied to an air base defense requirement process. Given that the Air Force motivation is to improve air base defense, not to take over theater AMD, such a clause seems a reasonable and small price to pay for the Air Force to obtain and control air base point air defenses.

The main argument against this COA is that it would require the Air Force leadership either to successfully advocate for additional funding and manpower billets or to shift resources from other programs to acquire these weapon systems and field new forces. The Air Force is better off under this COA only if the leadership is determined and able to devote significant resources to the mission.

COA 7: Push for Major Roles and Functions Overhaul

This is the most ambitious and likely riskiest of the six COAs, seeking a major reassessment of service R&F. The last major independent review of roles and missions was the 1995 CORM, roughly 24 years ago. Given the changes in the international security environment; new operational concepts, such as multidomain operations; and the pace of technological change, it is arguably time to look critically at the current assignment of responsibilities. Indeed, President Donald Trump's 2019 directive to DoD to create the Space Force may demand a reassessment of roles and missions across the board.[14] It also may be time to ask whether current R&F assignments among the services impede innovation and the acquisition of critical warfighting

---

[13] For example, in 1993, CSAF McPeak proposed to CSA Sullivan trading the CAS mission (and A-10 aircraft) with the Army in exchange for the theater air and missile defense mission (and Army Patriot and Hawk SAMs). See McPeak, 2017, pp. 313–315.

[14] Thanks to Paula Thornhill for this observation. For more on the Space Force, see Helene Cooper, "Trump Signs Order to Begin Creation of Space Force," *New York Times*, February 19, 2019.

capabilities. There appears to be some momentum in this direction already, and it may happen whether or not the Air Force advocates for it.

Whatever the arguments for a new CORM, this COA is not particularly promising as a means of addressing near-term shortfalls in air base defense. If a major R&F reassessment occurs, it will likely be a drawn out, high profile, and controversial affair with endless opportunities for distraction and interservice fighting. A future CORM might fix the air base defense problem under the rubric of some broader changes but could just as easily introduce new hurdles. In any event, air base defense is likely to be treated as a secondary or tertiary problem relative to other issues raised in a CORM study.

Air Force leaders may decide that it is time to push for a major reexamination of R&F based on other considerations. That may indeed be the right path for the institution. From the narrower perspective of air base defense, however, this COA courts considerable risk and offers no certainty that the air base defense issue will be resolved.

A Multipronged Strategy

Table 6.2 brings together these seven COAs, past or current examples of each, our assessment of their future prospects, and how they might contribute to a better alignment of service responsibilities for air base defense.

As discussed in the previous section, our assessment suggests that COAs 1, 3, and 6 are the most promising. COAs 4, 5, and 7 offer no immediate benefit and risk many potential costs in our assessment. COA 2, using JCIDS to generate requirements for the Army, should be pursued as a show of good faith and a sign of respect for existing joint processes. By itself, however, this COA is unlikely to be successful. As we noted earlier, COA 1, the pursuit of improvements within the Air Force's existing authorities, is essential for two reasons. First, the Air Force must make real investments to demonstrate its commitment to defend its air bases if it wants to be credible in arguing to own ground-based air defense. Second, the (largely) passive defenses that are under Air Force control are the most versatile defensive options and are generally much simpler, cheaper, and easier to acquire than other options. COA 3, using technology to overcome R&F paralysis, offers another avenue for the Air Force to demonstrate its commitment, this time via R&D investments directed at air base defense. Advanced technologies and innovative approaches (including partnering with other services, organizations, and the private sector) play to Air Force strengths, offer the potential for vastly improved defenses, and may be the most effective means to break down antiquated R&F lanes. COA 6, an MOU with the Army that would transfer air base defense to the Air Force, is the only COA that explicitly addresses the misalignment of service priorities and responsibilities.

Table 6.2. Potential Air Force Courses of Action to Overcome Roles and Functions Barriers

Courses of Action	Explanation/Precedents	Future Prospects
1. Pursue air base defense options that are clearly within Air Force purview	Passive defenses and Civil Engineer and Security Force manning and capabilities all belong to the Air Force	The Air Force must demonstrate its commitment to air base defense
2. Use JCIDS, PPBE, and DAS to address cruise missile defense shortfalls	In theory, ground-based air defense is a joint problem; combatant commanders should establish requirements	In practice, major programs are rarely funded without sustained advocacy from service leaders
3. Bypass R&F roadblocks through innovative use of technology	The Air Force has deployed Dronebuster (RF jammer) and THOR (HPM) for counter-SUAS	Directed energy may offer the Air Force a versatile point defense system that would avoid a fight over R&F
4. Propose new MOU: ground-based air defense of air bases is Army responsibility	In 1971, the Air Force proposed an MOU for air-based ground defense in Vietnam, JSA 8 on air-based ground defense	The Army has already been assigned this function, but defense of air bases is a low priority
5. Propose new MOU: ground-based air defense is joint Air Force–Army responsibility	In 1984, the MOU for air-based ground defense gave the Air Force authority to deploy such defense organically if Army unable	Neither the Army nor the Air Force improved SHORAD systems following the 1984 MOU
6. Propose new MOU: ground-based air defense of air bases is an Air Force responsibility	The Army's abrogation of JSA 8 in 2005 may offer a precedent	The Army might view this as a first step toward Air Force control of all land-based theater integrated AMD
7. Push for major R&F overall	The Air Force attempted to own all land-based theater-level AMD in the 1990s	Efforts to reform R&F might gain traction if they are comprehensive, but this is not without risks for the Air Force

No single COA appears to offer a high-probability path to success. Rather, a combination of COAs 1, 3, and 6 appears to offer the most robust strategy, bringing together unilateral Air Force self-help efforts, primarily infrastructure investments and new basing concepts (COA 1), multilateral R&D of advanced technologies (COA 3), and bilateral efforts to realign R&F (COA 6).

In Chapter 7, we present our findings and recommendations.

7. Findings and Recommendations

The growing vulnerability of U.S. air bases and other fixed facilities to enemy ballistic missile, cruise missile, and UAS attacks continues to gain prominence as a defense planning problem. However, these threats vary in difficulty and available solutions. Ballistic missile defense has proven to be extremely challenging, both technically and operationally, especially the cost imbalance between relatively cheap offensive ballistic missiles and expensive ballistic missile defense interceptors. The emerging problem of hypersonic missile defense only makes a difficult situation worse. The UAS threat, particularly that from swarms of small UASs, initially caught DoD off guard. In turn, the department and services have responded with energy and agility, especially in the use of advanced technologies, such as lasers and HPM weapons. Although these systems have yet to be fully developed and operationally deployed in the needed numbers, affordable solutions appear well in hand. In contrast, cruise missile defense is vastly easier than ballistic missile defense from a technical perspective, and, unlike with UAS defenses, highly capable cruise missile defense systems (e.g., NASAMS) are already in the field. Rather, the problem the Air Force and DoD face with cruise missile defense is institutional: a lack of leadership focus and determination to field solutions. Despite many contributing factors, a misalignment in service responsibilities dating back to the 1950s presents arguably the biggest barrier to improving air base defenses, particularly against cruise missiles. As discussed in Chapter 6, assigning air base air defense to the U.S. Army created a misalignment between the services' responsibilities and priorities.

This report is intended to inform Air Force deliberations on the future of air base defense, particularly with respect to the future of GBAD. The report contributes to these deliberations in several ways. The analysis of historical R&F here is the first to give air base defense the in-depth treatment it deserves, using previously unpublished primary source documents from the Air Force Historical Research Agency. Second, this is the first analysis to integrate threats, defensive options, constraints on R&F, and organizational strategies into a single framework. Finally, we believe the COAs recommended offer Air Force leaders a multidimensional strategy that is feasible and rapidly executable.

In this final chapter, we present research findings, make recommendations to Air Force leaders and planners, and offer some final thoughts on service R&F for air base defense.

Findings

Air Base Defense Has Been an Enduring Area of Disagreement and Frustration for the Army and the Air Force

The Air Force and Army have disagreed over their respective responsibilities for air base defense for most of the past 70 years. In the 1950s, they disputed the utility of AAA for air defense of SAC bases and disagreed over which SAM system was best for point air defense. In the 1960s and early 1970s, the Air Force repeatedly and unsuccessfully asked the Army to dedicate infantry forces to the ground defense of Air Force bases. The 1980s were the one period when the services did work cooperatively to establish and clarify their respective roles for base defense, but shortfalls had yet to be addressed when the Berlin Wall fell. Ground defense returned as a problem when insurgent attacks threatened air bases in Iraq and Afghanistan, and the Army found that it lacked the force structure to meet its commitments.

The dual-threat aspect of air base defense (defending from both ground and air threats) further complicates service R&F. Since the 1950s, documents on air base defense roles and missions distinguished between defending air bases from ground-based threats and defending the bases from air-based threats. Ironically, the Army has consistently displayed a preference and willingness to concede responsibility for defending air bases against ground threats but not against air threats. Today, after two decades of neglect, air defense of air bases has reemerged as a policy problem, yet unresolved disagreements about R&F remain a barrier to timely solutions.

Although Many Factors Are at Play, the Misalignment of Service Responsibilities and Priorities for Air Base Defense Is a Primary Factor of Enduring Shortfalls

As we discuss in Chapter 6, Army and Air Force responsibilities, stakes, priorities, and force structure are not well aligned for the air base defense function. The Army is responsible for providing point AMD for Air Force bases and other fixed facilities, but years of neglect have resulted in shortfalls, especially against cruise missiles. The Army leadership has understandably prioritized mobile short-range air defense for its maneuver units. While the Army is concurrently developing and fielding systems intended to address the threats to air bases and other priority point and fixed facilities (e.g., IFPC, Iron Dome, THAAD, Patriot), uncertainties remain, especially with regard to IFPC and Iron Dome. Moreover, it may be the case that the Air Force will continue to perceive the Army's commitment to defending air bases as low. Air bases are, after all, only one asset in a long list of priorities. Because the Air Force has much to lose if air bases are not properly defended, its sense of vulnerability may well persist, particularly in light of current capability and capacity shortfalls.

Difficulties in Validating Joint Warfighting Requirements, Army Resource Limitations, and Air Force Ambivalence Have Contributed to a Roadblock for Air Base Defense Roles and Functions

In theory, the JCIDS, PPBE, and DAS processes identify warfighter needs, which the services then address within their assigned functional areas. In practice, established joint requirements may be necessary but are not sufficient for new capabilities to be created, funded, and fielded. New technologies, concepts, and programs are typically initiated by the services; their advocacy is essential if a program is to survive the programming and budget gauntlet. In the case of GBAD for air bases, the joint staff cannot compel the Army to procure such systems if the service does not see the need. Army resource limitations are a second important factor.

After two decades of COIN combat operations, the Army is undergoing a difficult and costly reset to prepare for large-scale maneuver combat against a near peer. The modernization challenges are substantial, and it is quite reasonable for Army leaders to prioritize improvements in armored systems, long-range fires, EW, and forward air defenses over air base defense.[1]

The Air Force May Be Able to Bypass This Through Innovation Because Existing Roles and Functions Assignments Tend to Be Technology Centric

Although innovation advocates remain frustrated by what Defense Innovation Board chairman (and former Google chief executive officer) Eric Schmidt described as "a very bad system,"[2] progress has ensued. Technology partnering among the services, the Defense Advanced Research Projects Agency, other agencies, and the private sector is rapidly advancing key air base defense technologies. For example, all four services are urgently pursuing defenses against small UASs tailored to meet each service's particular tactical requirements. Multiple services are developing directed-energy systems, using HPM systems and lasers, which appear to be the most promising. None of the services seems at all concerned about R&F as they rush prototype systems to the field for testing.

This suggests that the Air Force may be able to develop its own GBAD if the systems use advanced technologies rather than legacy technologies and designs. This finding is somewhat speculative; much more research will need to be done to determine whether this is so. One possibility is that innovation-centered partnering creates new cultural norms for participants, primarily a low tolerance for bureaucratic rules, traditions, and structure. The culture of innovation emphasizes breaking down any and all barriers to creative problem solving, experimentation, and progress. These new cultural norms may be the most effective, and perhaps the only, way to move past antiquated approaches to service R&F.

---

[1] The Air Force, too, must make such priority choices about the weapons and systems it fields. These may not align with Army preferences, especially when it comes to CAS and airlift.

[2] Carden Cordell, "What's Impeding the DOD's Push for Innovation? Turns Out, a Lot," Fedscoop website, April 17, 2018.

The Most Robust Strategy to Improve Air Base Defenses Would Pursue Parallel Lines of Effort

Our analysis of seven alternative COAs concluded that no single COA offered a high probability of success. Two COAs (both new MOUs) were discarded because the potential payoffs appeared quite small relative to the effort required. Another was discarded because it would be high risk and unlikely to improve air base defenses in the near term. One COA (using JCIDS) should be pursued but is unlikely to result in improved ground-based air defense capabilities unless there is strong support for them from Army leaders. The remaining three COAs are each worth pursuing on their own merits, but no one COA is sufficient. Rather, a combination of the following three appears most robust: (1) internal Air Force efforts to prioritize integrated air base defense programs, (2) expanded use of innovation hubs and technology partners to break down R&F barriers, and (3) high-level engagement with Army leaders to resolve long-standing misalignments of R&F for air base defense.

Recommendations

These research findings suggest that the Air Force should pursue parallel lines of effort to improve air base defenses. Specifically, we recommend that the Air Force:

- **Demonstrate institutional commitment to air base defense by increasing the number of personnel dedicated to air base resiliency, significantly expanding the scope and pace of defensive programs, and more visibly advocating for air base defense.** No R&F barriers prevent the Air Force from rapidly improving the resiliency of air bases to attack. Increasing the number and size of Security Force units, constructing hardened structures (e.g., for aircraft, command posts, personnel), purchasing of deployable protective shelters and fuel bladders, and developing air base recovery capabilities are all relatively cheap and simple programs that can be executed quickly. They make measurable contributions to air base defense and are tangible proof of Air Force determination to meet this challenge.

- **Embrace the Air Force culture of innovation to break down R&F barriers.** The AFWERX innovation hub is only the latest example of the Air Force's institutional commitment to pushing technological and operational frontiers. Through partnering with DoD and the other services, the Air Force is advancing promising air base defense technologies, creating a cross-service culture of innovation, and bypassing some R&F barriers. The cross-service partnerships, by themselves, have the potential to create a new community of innovators. This community would transcend all the services, comprising those unwilling to be bound by narrow R&F. Additionally, the development and fielding of prototype weapons using new technologies may allow the Air Force to create facts on the ground—real warfighting capabilities deployed and used in combat. Fielding prototype systems does not automatically circumvent R&F but may offer a means of avoiding related paralysis associated with more-traditional planning, programming, and budgeting processes.

- **Propose a new MOU with the Army to establish ground-based air defense of air bases as an Air Force responsibility, correcting the misalignment of service responsibilities and priorities.** This recommendation depends greatly on the relationship between CSAF and CSA. If it is similar to that of Generals Gabriel and Wickham in the 1980s, this MOU might be only one of many seeking to improve cooperation between the two services. If, however, there is a lack of trust or protective attitude toward the mission on the part of the CSA, this COA may be dead on arrival. On the merits of the idea, it seems like an easy sell, assuming that the MOU is sensitive to Army concerns (i.e., an Air Force attempt to control all AMD) and limits Air Force ground-based air defense to air base defense only. The Army has shown a willingness to give up air base defense responsibilities in the event of resource shortfalls, as it did during Operation Iraqi Freedom. The modernization challenge the Army faces today is arguably even more consequential than Operation Iraqi Freedom force structure shortfalls. Giving up the air base air defense mission would allow the Army to focus on more-pressing concerns. Finally, proposing such an MOU is further evidence of the Air Force's intention to own this problem and would complement the internal Air Force and technology partnering initiatives.

Final Thoughts

Just as airmen have been the most outspoken advocates for airpower as a strategic instrument, so too must airmen lead the way on air base defense. No other organization—not OSD, the Joint Staff, the combatant commands, MDA, or Congress—can take the place of airmen in conceptualizing the problem and articulating solutions. The level of commitment to air base defense these other organizations have shown will always be some fraction of what airmen have shown. If airmen are ambivalent in their commitment, these other actors will be even more so.

Moreover, given the centrality of air bases in generating airpower, it may well be that airmen will always perceive insufficient concern for protecting runways and airfields on the part of other services or even the combatant commands. It therefore is understandable that the Air Force encourage the other services and the AMD community to invest more in air base defenses. The question that remains is whether the Air Force is prepared to do so itself.

Although senior airmen have increasingly emphasized the importance of base defense and although many promising concepts and technologies have been tested and developed, the Air Force would be hard pressed to show concrete evidence (e.g., in funding and fielded capabilities) that the institution is serious about the problem. Neither have airmen brought the intellectual energy they previously devoted to offensive air campaigns to the problem of air base defense. There are many tactical- and operational-level innovations in progress, but the Air Force has yet to integrate air base defense into airpower narratives and has not created a concept for truly integrated air base defense against the full range of threats, from enemy commandos to hypersonic missiles. If Air Force air base defense efforts are to gain traction with outside audiences, the service will first need to address these shortfalls in airpower theory, strategy, and operational concepts.

Bibliography

Abramson, Rudy, "Weinberger Kills Anti-Aircraft Gun: After $1.8 Billion, He Says Sgt. York Is Ineffective, Not Worth Further Cost," *Los Angeles Times*, August 28, 1985.

Acton, James M., *Silver Bullet? Asking the Right Questions About Conventional Prompt Global Strike*, Washington, D.C.: Carnegie Endowment for International Peace, 2013.

AeroVironment, "Switchblade," website, undated. As of June 5, 2019:
https://www.avinc.com/uas/view/switchblade

AFH—*See* Air Force Handbook.

AFM—*See* Air Force Manual.

AFTTP—*See* Air Force Tactics, Techniques, and Procedures.

Air Defense Artillery Association, "1991 Air Defense Artillery Yearbook," *Air Defense Artillery*, Spring 1991.

Air Force Handbook 10-222, Vol. 1, *Civil Engineer Bare Base Development*, Washington, D.C.: Department of the Air Force, January 23, 2012.

Air Force Handbook 10-222, Vol. 14, *Civil Engineer Guide to Fighting Positions, Shelters, Obstacles and Revetments*, Washington, D.C.: Department of the Air Force, August 1, 2008.

Air Force Manual 31-201v1, *Security Forces History*, Washington, D.C.: Department of the Air Force, August 9, 2010.

Air Force Manual 207-1, *Doctrine and Requirements for Security of Air Force Weapons Systems*, Washington, D.C.: Department of the Air Force, June 10, 1964.

Air Force Pamphlet 10-219, Vol. 2, *Civil Engineer Disaster and Attack Preparations*, Washington, D.C.: Department of the Air Force, June 9, 2008.

Air Force Tactics, Techniques, and Procedures 3-10.1, *Integrated Base Defense (IBD)*, August 20, 2004.

Air Force Tactics, Techniques, and Procedures 31-1, *Integrated Defense*, 2007.

Air Force Tactics, Techniques, and Procedures 3-32.34V3, *Civil Engineer Expeditionary Force Protection*, March 1, 2016.

Air Land Sea Application Center, "ALSA Roadshow," briefing slides, September 5, 2019. As of May 6, 2020:
https://www.alsa.mil/Portals/9/Documents/roadshow.pdf

Almodovar, Gabriel, Daniel P. Allmacher, Morgan P. Ames III, and Chad Davies, "Joint Integrated Air and Missile Defense: Simplifying an Increasingly Complex Problem," *Joint Forces Quarterly*, Vol. 88, 1st Quarter 2018, pp. 78–84. As of March 6, 2020: https://ndupress.ndu.edu/Portals/68/Documents/jfq/jfq-88/ jfq-88_78-84_Almodovar-et-al.pdf?ver=2018-01-09-102341-613

ALSA—*See* Air Land Sea Application Center.

Andrade, Dale, "Westmoreland Was Right: Learning the Wrong Lessons from the Vietnam War," *Small Wars and Insurgencies*, Vol. 19, No. 2, June 2008.

Arakaki, Leatrice R., and John R. Kuborn, *7 December 1941: The Air Force Story*, Hickam AFB, Hawaii: Pacific Air Forces Office of History, 1991.

Army Tactics, Techniques, and Procedures 3-34.39, *Camouflage, Concealment and Decoys*, Washington, D.C.: Headquarters, Department of the Army, November 2010.

Armstrong, Michael J., "The Effectiveness of Rocket Attacks and Defenses in Israel," *Journal of Global Security Studies,* Vol. 3, No. 2, April 2018, pp. 113–132. As of March 6, 2020: https://academic.oup.com/jogss/article/3/2/113/4964794

Arquilla, John, and David Ronfeldt, *Swarming and the Future of Conflict*, Santa Monica, Calif.: RAND Corporation, DB-311-OSD, 2000. As of March 11, 2020: https://www.rand.org/pubs/documented_briefings/DB311.html

Assistant Vice Chief of Staff of the Air Force, letter to TAC Lt. General Robbins, Washington, D.C.: Department of the Air Force, Office of the Chief of Staff, April 12, 1971. (Department of the Air Force Archives at Maxwell AFB, Ala.)

Atherton, Kelsey, "As Counter-UAS Gains Ground, Swarm Threat Looms," *Aviation Week and Space Technology*, March 26–April 8, 2018, pp. 36–37.

"Avenger Low Level Air Defence System, USA," Army Technology website, undated. As of July 16, 2019: https://www.army-technology.com/projects/avenger/

Barrie, Douglas, "Kh-101 Missile Test Highlights Russian Bomber Firepower," *Military Balance Blog*, IISS, February 8, 2019. As of May 10, 2019: https://www.iiss.org/blogs/military-balance/2019/02/russian-bomber-firepower

Bartsch, William H., *December 8, 1941: MacArthur's Pearl Harbor*, College Station, Tex.: Texas A&M University Press, 2003.

Beers, Clay, Gordon Miller, Robert Taradash, and Parker Wright, *Zone Defense: A Case for Distinct Service Roles and Missions*, Washington, D.C.: Center for New American Security, January 2014.

Bell, Raymond E., "To Protect an Air Base," *Air Power Journal*, Vol. 3, No. 3, Fall 1989, pp. 4–19.

Bender, Bryan, "USACOM Mission May Be Adjusted: CORM to Recommend New 'Functional' Command for Joint Force Training," *Inside the Army*, Vol. 7, No. 20, May 22, 1995, pp. 1, 10–11.

Benson, Lawrence R., *USAF Aircraft Basing in Europe, North Africa, and the Middle East: 1945–1980*, Ramstein AB, Germany: Headquarters, U.S. Air Forces in Europe, 1981.

Berger, Carl, ed., *The United States Air Force in Southeast Asia: 1961–1973*, Washington, D.C.: Office of Air Force History, 1977. As of March 6, 2020:
https://apps.dtic.mil/dtic/tr/fulltext/u2/a045012.pdf

Bergerud, Eric M., *Fire in the Sky: The Air War in the South Pacific*, Boulder, Colo.: Westview Press, 2001.

Berhow, Mark, *U.S. Strategic and Defensive Missile Systems: 1950–2004*, Oxford, UK: Osprey Publishing, 2005.

Betts, Richard K., ed., *Cruise Missiles: Technology, Strategy, Politics*, Washington, D.C.: Brookings Institution, 1981.

Bingham, Price T., "Fighting from the Air Base," *Air Power Journal*, Summer 1987, pp. 32–41.

Bonomo, James, Giacomo Bergamo, David R. Frelinger, John Gordon IV, and Brian A. Jackson, *Stealing the Sword: Limiting Terrorist Use of Advanced Conventional Weapons*, Santa Monica, Calif.: RAND Corporation, MG-510-DHS, 2007. As of June 5, 2019:
https://www.rand.org/pubs/monographs/MG510.html

Bower, Laura, "Defending Europe's Skies: The 32nd AADCOM Today," *Air Defense Artillery*, Summer 1983, pp. 30–32.

Bowie, Christopher J., *The Anti-Access Threat and Theater Air Bases*, Washington, D.C.: Center for Strategic and Budgetary Assessments, 2002.

Bowie, Christopher J., "The Lessons of Salty Demo," *Air Force Magazine*, March 2009. As of March 11, 2020:
https://www.airforcemag.com/article/0309salty/

Briar, David P., "Sharpening the Eagles' Talons: Assessing Advance in Air Base Defense Doctrine," in Caudill, 2014, pp. 263–280.

Brouse, Steven M., "Congress Revives Roles and Missions Debate," *Air Defense Artillery*, November–December 1994, p. 22.

Brumfiel, Geoff, "Nations Rush Ahead with Hypersonic Weapons amid Arms Race Fear," National Public Radio, October 23, 2018. As of March 11, 2020: https://www.npr.org/2018/10/23/659602274/amid-arms-race-fears-the-u-s-russia-and-china-are-racing-ahead-with-a-new-missil

Buchannan, Walter E., III, Bradley D. Spacy, Chris Bargery, Glen E. Christensen, and Armand "Beaux" Lyons, "Outside the Wire: Recollections of Operations Desert Safeside and TF 1041," in Caudill, 2019, pp. 137–208.

Buss, L. H., Lloyd H. Cornett, Jr., Elsie L. Joerling, and Derril E. Howell, *Continental Air Defense Command Historical Summary: July 1956–June 1957*, Colorado Springs, Colo.: Continental Air Defense Command Office of History, September 15, 1957.

Castle, Ian, *The Zeppelin Base Raids: Germany 1914*, Oxford, UK: Osprey Publishing, 2011.

Caston, Lauren, Robert S. Leonard, Christopher A. Mouton, Chad J. R. Ohlandt, S. Craig Moore, Raymond E. Conley, and Glenn Buchan, *The Future of the U.S. Intercontinental Ballistic Missile Force*, Santa Monica, Calif.: RAND Corporation, MG-1210-AF, 2014. As of December 16, 2019: https://www.rand.org/pubs/monographs/MG1210.html

Caudill, Shannon, ed., *Defending Air Bases in an Age of Insurgency*, Maxwell AFB, Ala.: Air University Press, 2014.

Caudill, Shannon, ed., *Defending Air Bases in an Age of Insurgency*, Vol. II, Maxwell AFB, Ala.: Air University Press, 2019.

Center for Arms Control and Non-Proliferation, "Ballistic vs. Cruise Missiles," fact sheet, undated. As of April 22, 2019: https://armscontrolcenter.org/wp-content/uploads/2017/04/Ballistic-vs.-Cruise-Missiles-Fact-Sheet.pdf

Chairman of the Joint Chiefs of Staff Instruction 5123.01H, *Charter of the Joint Requirements Oversight Council (JROC) and Implementation of the Joint Capabilities Integration and Development System (JCIDS)*, Washington, D.C.: Joint Staff, August 31, 2018. As of December 15, 2019: https://www.jcs.mil/Portals/36/Documents/Library/Instructions/CJCSI%205123.01H.pdf?ver=2018-10-26-163922-137

CHECO—*See* Contemporary Historical Examination of Current Operations.

Chief of Staff of the Air Force, AFCCS 76672, 0113502, message to Commander in Chief, U.S. Pacific Air Forces, September 1965.

Christensen, Glen E., *Air Base Defense in the Twenty-First Century*, Fort Leavenworth, Kan.: School of Advanced Military Studies, 2007.

Clark, Bryan, and Mark Gunzinger, *Winning the Airwaves: Regaining America's Dominance in the Electromagnetic Spectrum*, Washington, D.C.: Center for Strategic and Budgetary Assessments, 2017.

Clark, Colin, "Air Force Launches Electronic Warfare Roadmap: EMS ECCT 2.0," Breaking Defense website, April 24, 2019. As of June 30, 2019:
https://breakingdefense.com/2019/04/air-force-launches-electronic-warfare-roadmap-ems-ecct-2-0/

Coffed, Jeff, *The Threat of GPS Jamming: The Risk to an Information Utility*, Melbourne, Fla.: Harris Corporation, 2016. As of June 30, 2019:
https://www.harris.com/sites/default/files/downloads/solutions/d0783-0063_threatofgpsjamming_v2_mv.pdf

Cohen, Rachel, "The Drone Zappers," *Air Force Magazine*, March 22, 2019a. As of March 6, 2020:
https://www.airforcemag.com/article/the-drone-zappers/

Cohen, Rachel, "Strategic Air Bases Receive First Counter-UAS Systems," *Air Force Magazine*, July 1, 2019b. As of March 6, 2020:
https://www.airforcemag.com/strategic-air-bases-receive-first-counter-uas-systems/

Cohen, Raphael S., *The History and Politics of Defense Reviews*, Santa Monica, Calif.: RAND Corporation, RR-2278-AF, 2018. As of March 5, 2020:
https://www.rand.org/pubs/research_reports/RR2278.html

Cole, William, "USS Preble to Be First Destroyer Equipped with Laser Defense System," *Honolulu Star-Advertiser*, May 29, 2019. As of March 11, 2020:
https://www.military.com/daily-news/2019/05/29/uss-preble-be-first-destroyer-equipped-laser-defense-system.html

Commission on Roles and Missions of the Armed Forces, *Directions for Defense*, Washington, D.C.: U.S. Department of Defense, 1995. As of March 6, 2020:
https://apps.dtic.mil/dtic/tr/fulltext/u2/a295228.pdf

Comptroller General of the United States, *Evaluation of the Decision to Begin Production of the Roland Missile System*, Washington, D.C.: General Accounting Office, August 17, 1979.

"Congressman Accuses Pentagon of Buying 'Unnecessary' Missile," *New York Times*, June 25, 1979.

Contemporary Historical Examination of Current Operations Office, *Historical Background to Viet Cong Mortar Attack on Bien Hoa: 1 November 1964*, Honolulu, Hawaii: Headquarters 2nd Air Division, November 9, 1964.

Cooling, Benjamin Franklin, ed., *Case Studies in the Achievement of Air Superiority*, Washington, D.C.: Air Force History and Museums Program, 1994.

Cooper, Helene, "Trump Signs Order to Begin Creation of Space Force," *New York Times*, February 19, 2019.

Cordell, Carden, "What's Impeding the DOD's Push for Innovation? Turns Out, a Lot," Fedscoop website, April 17, 2018. As of August 16, 2019: https://www.fedscoop.com/whats-impeding-dods-push-innovation-turns-lot/

Cordesman, Anthony H., and Abraham R. Wagner, *The Lessons of Modern War*, Vol. III: *The Afghan and Falklands Conflict*, Boulder, Colo.: Westview Press, 1990.

Corey, Craig R., "The Air Force's Misconception of Integrated Air and Missile Defense," *Air and Space Power Journal*, Vol. 31, No. 4, Winter 2017, pp. 81–90. As of March 6, 2020: https://www.airuniversity.af.edu/Portals/10/ASPJ/journals/Volume-31_Issue-4/V-Corey.pdf

Correll, John T., "Air Defense from the Ground Up," *Air Force Magazine*, Vol. 66, No. 7, July 1983, pp. 37–43. As of March 6, 2020: https://www.airforcemag.com/issue/1983-07/

Correll, John T., "A New Look at Roles and Missions," *Air Force Magazine*, Vol. 91, No. 11, November 2008, pp. 50–54.

Crabtree, James D., *On Air Defense*, Westport, Conn.: Praeger Publishers, 1994.

Craven, Wesley F., and James L. Cate, *The Army Air Forces in World War II*, Vol. VI: *Men and Planes*, Washington, D.C.: Office of Air Force History, [1955] 1983.

Cravens, James J., "Intercept Point," *Air Defense Artillery*, July–August 1995, p. 1.

CSAF—*See* Chief of Staff of the Air Force.

CSIS Missile Defense Project "Kh-101/Kh-102," Missile Threat webpage, June 15, 2018. As of April 23, 2020: https://missilethreat.csis.org/missile/kh-101-kh-102/

Curtis, Nicole, "C-RAM Task Force: Air Defense Artillery School Continues Quest for Ultimate Counter-Rockets, Artillery and Mortars Interceptor," *Air Defense Artillery*, July–September 2005, p. 7.

Davis, Lynn E., Michael J. McNerney, James Chow, Thomas Hamilton, Sarah Harting, and Daniel Byman, *Armed and Dangerous? UAVs and U.S. Security*, Santa Monica, Calif.: RAND Corporation, RR-449-RC, 2014. As of March 5, 2020: https://www.rand.org/pubs/research_reports/RR449.html

Davis, Richard G., *The 31 Initiatives: A Study in Air Force–Army Cooperation*, Washington, D.C.: Office of Air Force History, 1987. As of March 18, 2020: https://history.army.mil/html/books/106/106-1/CMH_Pub_106-1.pdf

Davis, Vincent, *Postwar Defense Policy and the U.S. Navy, 1943–1946*, Chapel Hill, N.C.: University of North Carolina Press, 1966.

Defense Advanced Research Projects Agency, "Long Range Anti-Ship Missile," webpage. As of August 14, 2019: https://www.darpa.mil/about-us/long-range-anti-ship-missile

Defense Intelligence Ballistic Missile Analysis Committee, *2017 Ballistic and Cruise Missile Threat*, Wright-Patterson AFB, Ohio: NASIC Public Affairs Office, June 30, 2017. As of May 10, 2019: https://www.nasic.af.mil/About-Us/Fact-Sheets/Article/1235024/2017-ballistic-and-cruise-missile-threat-report/

De Longe, Merrill E., *Modern Airfield Planning and Concealment*, New York: Pitman Publishing Company, 1943.

"Demise of the SGT. York: Model Turns to Dud," *New York Times*, December 2, 1985.

DeMott, John S., "Son of the Sergeant York," *Time*, June 24, 2001. As of July 23, 2019: http://content.time.com/time/magazine/article/0,9171,144737,00.html

Dennison, John W., and Melvin F. Porter, *Local Base Defense in RVN: January 1969–June 1971*, Project CHECO Southeast Asia Report, Honolulu, Hawaii: Headquarters PACAF, September 14, 1971.

Department of the Army, Justification of Estimates for Fiscal Year 1985, Submitted to Congress February 1984, Procurement Programs: Aircraft, Missiles, Weapons & Tracked Combat Vehicles, Ammunition, Other Procurement, and Construction Programs, Washington, D.C.: Office of the Deputy Chief of Staff for Research, Development, and Acquisition, February 1984.

Department of the Army and Department of the Air Force, "Memorandum of Understanding on United States Army (USA)/United States Air Force (USAF) Responsibilities for Air Base Air Defense," Washington, D.C.: Headquarters, U.S. Army and Headquarters, U.S. Air Force, July 13, 1984, as reproduced in Davis, 1987, pp. 120–124.

———, "Joint Service Agreement on United States Army (USA)/United States Air Force (USAF) Agreement for the Ground Defense of Air Force Bases and Installations," Washington, D.C.: Headquarters, U.S. Army and Headquarters, U.S. Air Force, April 25, 1985, as reproduced in Davis, 1987, pp. 125–131.

———, *Abrogation of Joint Security Agreement 8*, Washington, D.C.: Department of Defense, March 24, 2005.

Department of Defense 6055.09-STD, *DoD Ammunition and Explosives Safety Standards*, Washington, D.C.: Office of the Deputy Under Secretary of Defense (Installations and Environment), February 29, 2008, Incorporating Change 2, August 21, 2009.

Department of Defense Directive 5100.01, *Functions of the Department of Defense and Its Major Components*, Washington, D.C.: U.S. Department of Defense, 2010.

Deptula, David A., *Revisiting the Roles and Missions of the Armed Forces,* statement before the Senate Armed Services Committee, Washington, D.C.: Mitchell Institute for Aerospace Studies, November 5, 2015. As of March 6, 2020:
https://www.armed-services.senate.gov/imo/media/doc/Deptula_11-05-15.pdf

Deptula, David, "America's Air Superiority Crisis," Breaking Defense website, July 12, 2017. As of June 30, 2019:
https://breakingdefense.com/2017/07/americas-air-superiority-crisis/

Director of Central Intelligence, *Warsaw Pact Nonnuclear Threat to NATO Airbases in Central Europe: National Intelligence Estimate*, Washington, D.C.: Central Intelligence Agency, NIE 11/20-6-84, October 25, 1984. As of May 10, 2019:
http://www.foia.cia.gov/sites/default/files/document_conversions/89801/DOC_0000278545.pdf

Don, Bruce W., Donald E. Lewis, Robert M. Paulson, and Willis W. Ware, *Survivability Issues and USAFE Policy*, Santa Monica, Calif.: RAND Corporation, N-2579-AF, 1988. As of March 6, 2020:
https://www.rand.org/pubs/notes/N2579.html

Douhet, Giulio, *The Command of the Air*, trans. Dino Ferrari, Washington, D.C.: Office of Air Force History, [1921, Italian] 1983.

"Dubai Airport Grounds Flights Due to 'Drone Activity,'" BBC News website, September 28, 2016. As of March 11, 2020:
https://www.bbc.com/news/world-middle-east-37493404

Eckstein, Megan, "Marines' Anti-Drone Defense System Moving Towards Testing, Fielding Decision by End of Year," USNI News website, March 11, 2019. As of July 1, 2019:
https://news.usni.org/2019/03/11/marines-anti-drone-defense-system-working-towards-testing-fielding-decision-by-end-of-year#more-41780

Edwards, Sean J. A., *Swarming and the Future of Warfare*, dissertation, Pardee RAND Graduate School, Santa Monica, Calif.: RAND Corporation, RGSD-189, 2005. As of March 26, 2020:
https://www.rand.org/pubs/rgs_dissertations/RGSD189.html

Eilon, Lindsey, Jack Lyon, Jason Zaborski, and Robin Rosenberger, *White Paper: Evolution of Department of Defense Directive 5100.01 'Functions of the Department of Defense and Its Major Components,'* Washington, D.C.: U.S. Department of Defense, January 2014. As of July 23, 2019:
https://cmo.defense.gov/Portals/47/Documents/PDSD/DoDD5100.01_WhitePaper.pdf

Elkins, Walter, "32nd Army Air Defense Command," U.S. Army in Germany website, undated. As of March 10, 2020:
https://www.usarmygermany.com/Sont.htm?https&&&www.usarmygermany.com/Units/Air%20Defense/USAREUR_32nd%20AADCOM.htm

Elsam, Mike B., *Brassey's Air Power: Aircraft, Weapons Systems and Technology Series*, Vol. 7: *Air Defence*, London: Brassey's Defence Publishers, 1989.

Emerson, Donald E., *USAFE Airbase Operations in a Wartime Environment*, Santa Monica, Calif.: RAND Corporation, P-6810, 1982. As of March 6, 2020:
https://www.rand.org/pubs/papers/P6810.html

Evans, Ryan, "Call for Articles: The Military Roles and Missions Analysis That America Deserves," War on the Rocks website, August 15, 2018. As of July 23, 2019:
https://warontherocks.com/2018/08/call-for-papers-the-military-roles-and-missions-analysis-that-america-deserves/

Executive Order 9877, Functions of the Armed Forces, July 26, 1947.

Fabian, Billy, Mark Gunzinger, Jan van Tol, Jacob Cohn, and Gillian Evans, *Strengthening the Defense of NATO's Eastern Frontier*, Washington, D.C.: Center for Strategic and Budgetary Assessments, 2019.

Farley, Robert, "AirLand Battle: The Army's Cold War Plan to Crush Russia (That Ended Up Crushing Iraq)," *National Interest*, August 1, 2018. As of July 30, 2019:
https://nationalinterest.org/blog/buzz/airland-battle-armys-cold-war-plan-crush-russia-ended-crushing-iraq-27477

Field Manual 3-01.11, *Air Defense Artillery Reference Handbook*, Washington, D.C.: Headquarters, Department of the Army, October 23, 2007.

Field Manual 3-22.90, *Mortars*, Washington, D.C.: Headquarters, Department of the Army, December 2007.

Field Manual 5-430-0-2 and Air Force Joint Pamphlet 32-8013, Vol. II, *Planning and Design of Roads, Airfields, and Heliports in the Theater of Operations—Airfield and Heliport Designs*, Washington, D.C.: Headquarters, Department of the Army and Department of the Air Force, September 1994.

Field Manual 100-5, *Operations*, Washington, D.C.: Headquarters, Department of the Army, 1982.

FM—*See* Field Manual.

Fogleman, Ronald R., and Sheila E. Widnall, *Global Engagement: A Vision of the 21st Century Air Force*, Washington, D.C.: Department of the Air Force, 1996. As of March 6, 2020: https://apps.dtic.mil/dtic/tr/fulltext/u2/a318235.pdf

Forrestal, James V., "Functions of the Armed Forces and the Joint Chiefs of Staff," appendix to "Note by the Secretaries to the Joint Chiefs of Staff," Washington, D.C.: U.S. Department of Defense, April 21, 1948. As of April 24, 2020: http://cgsc.cdmhost.com/cdm/singleitem/collection/p4013coll11/id/729/rec/1

Fox, Roger P., "Air Base Defense: An Appraisal," *Aerospace Commentary*, Vol. V, No. 1, Winter 1972.

Fox, Roger P., *Air Base Defense in the Republic of Vietnam 1961–1973*, Washington, D.C.: Office of Air Force History, 1979. As of March 6, 2020: https://media.defense.gov/2010/Sep/21/2001330253/-1/-1/0/AFD-100921-023.pdf

Frantzman, Seth J., "Can Iron Dome Cut It for Indirect Fire Protection? U.S. Army Is Buying a Couple Systems to Find Out," Defense News website, February 6, 2019. As of May 22, 2019: https://www.defensenews.com/global/mideast-africa/2019/02/06/can-iron-dome-cut-it-for-indirect-fire-protection-us-army-is-buying-a-couple-systems-to-find-out/

Freedberg, Sydney J., Jr., "The Limits of Lasers: Missile Defense at Speed of Light," Breaking Defense website, May 30, 2014. As of June 12, 2019: https://breakingdefense.com/2014/05/the-limits-of-lasers-missile-defense-at-speed-of-light/

Freedberg, Sydney J., Jr., "Army Reboots Cruise Missile Defense: IFPC & Iron Dome," Breaking Defense, March 11, 2019a. As of March 10, 2020: https://breakingdefense.com/2019/03/army-reboots-cruise-missile-defense-ifpc-iron-dome/

Freedberg, Sydney J., Jr., "New Army Laser Could Kill Cruise Missiles," Breaking Defense website, August 5, 2019b. As of August 13, 2019: https://breakingdefense.com/2019/08/newest-army-laser-could-kill-cruise-missiles/

Futrell, Robert F., *The United States Air Force in Southeast Asia: The Advisory Years to 1965*, Washington, D.C.: Office of Air Force History, 1981.

Futrell, Robert F., *Ideas, Concepts, Doctrine: Basic Thinking in the United States Air Force*, Vol. II: *1961–1984*, Maxwell Air Force Base, Ala.: Air University Press, 1989.

Gambrell, Jon, "How Yemen's Rebels Increasingly Deploy Drones," Defense News website, May 21, 2019. As of May 27, 2019:
https://www.defensenews.com/unmanned/2019/05/21/how-yemens-rebels-increasingly-deploy-drones/

GAO—See U.S. Government Accountability Office.

Gao, Charlie, "Why China Dominates the Short-Range Air Defense Game," *National Interest*, April 8, 2018.

Garrett, Joseph G., and Stephen M. Beatty, "My Thoughts on the Inactivation," *Air Defense Artillery*, July–August 1995, p. 4.

Genter, Thomas M., "SLAMRAAM Is Coming to a Theater Near You!" *Air Defense Artillery*, July–September 2005, pp. 10–12.

Gibbons-Neff, Thomas, Eric Schmitt, Charlie Savage, and Helene Cooper, "Chaos as Militants Overran Airfield, Killing 3 Americans in Kenya," *New York Times*, January 22, 2020.

Gibbons-Neff, Thomas, "More American Troops Sustain Brain Injuries from Iran Missile Strike in Iraq," *New York Times*, January 30, 2020.

Gons, Eric Stephen, *Access Challenges and Implications for Airpower in the Western Pacific*, dissertation, Pardee RAND Graduate School, Santa Monica, Calif.: RAND Corporation, RGSD-267, 2011. As of March 12, 2020:
https://www.rand.org/pubs/rgs_dissertations/RGSD267.html

Gormley, Dennis M., *Dealing with the Threat of Cruise Missiles*, New York: Oxford University Press for the International Institute for Strategic Studies, 2001.

Gormley, Dennis M., Andrew S. Erickson, and Jingdong Yuan, *A Low-Visibility Force Multiplier: Assessing China's Cruise Missile Ambitions*, Washington, D.C.: National Defense University Press, 2014.

Gould, Joe, "Guided-Bomb Makers Anticipate GPS Jammers," Defense News website, May 31, 2015. As of June 28, 2019:
https://www.defensenews.com/air/2015/05/31/guided-bomb-makers-anticipate-gps-jammers/

Grady, John, "DoD Official: U.S. Needs to Develop New Counters to Future Hypersonic Missiles," U.S. Naval Institute News website, November 16, 2018. As of March 11, 2020:
https://news.usni.org/2018/11/16/dod-official-u-s-needs-develop-new-counters-future-hypersonic-missiles

Grant, Rebecca, "Safeside in the Desert," *Air Force Magazine*, May 6, 2008, p. 47. As of March 9, 2020:
https://www.airforcemag.com/article/0207desert/

Groll, Elias, "Russia Is Tricking GPS to Protect Putin," *Foreign Policy*, April 3, 2019. As of June 28, 2019:
https://foreignpolicy.com/2019/04/03/russia-is-tricking-gps-to-protect-putin/

Grynkewich, Alex, "An Operational Imperative: The Future of Air Superiority," *Mitchell Institute Policy Papers*, Vol. 7, July 2017. As of June 30, 2019:
http://docs.wixstatic.com/ugd/a2dd91_4638a708c6a14f18ab9922d1c07930b3.pdf

Gunzinger, Mark, and Bryan Clark, *Winning the Salvo Competition: Rebalancing America's Air and Missile Defenses*, Washington, D.C.: Center for Strategic and Budgetary Assessments, 2016.

Gunzinger, Mark, and Christopher Dougherty, *Outside-In: Operating from Range to Defeat Iran's Anti-Access and Area-Denial Threats*, Washington, D.C.: Center for Strategic and Budgetary Assessments, January 17, 2012.

Hagen, Jeff, "Potential Effects of Chinese Aerospace Capabilities on U.S. Air Force Operations," testimony before the U.S.-China Economic and Security Review Commission, Santa Monica, Calif.: RAND Corporation, CT-347, May 20, 2010. As of March 9, 2020:
https://www.rand.org/pubs/testimonies/CT347.html

Hagen, Jeff, Forrest E. Morgan, Jacob L. Heim, and Matthew Carroll, *The Foundations of Operational Resilience—Assessing the Ability to Operate in an Anti-Access/Area Denial (A2/AD) Environment: The Analytical Framework, Lexicon, and Characteristics of the Operational Resilience Analysis Model (ORAM)*, Santa Monica, Calif.: RAND Corporation, RR-1265-AF, 2016a. As of March 6, 2020:
https://www.rand.org/pubs/research_reports/RR1265.html

Hagen, Jeff, David A. Blancett, Michael Bohnert, Shuo-Ju Chou, Amado Cordova, Thomas Hamilton, Alexander C. Hou, Sherrill Lingel, Colin Ludwig, Christopher Lynch, Muharrem Mane, Nicholas A. O'Donoughue, Daniel M. Norton, Ravi Rajan, and William Stanley, *Needs, Effectiveness, and Gap Assessment of Key A-10C Missions: An Overview of Findings*, Santa Monica, Calif.: RAND Corporation, RR-1724/1-AF, 2016b. As of March 10, 2020:
https://www.rand.org/pubs/research_reports/RR1724z1.html

Halliday, John M., *Tactical Dispersal of Fighter Aircraft: Risk, Uncertainty, and Policy Recommendations*, Santa Monica, Calif.: RAND Corporation, N-2443-AF, 1987. As of March 6, 2020:
https://www.rand.org/pubs/notes/N2443.html

Hallion, Richard P., *Control of the Air: The Enduring Requirement*, Bolling AFB, D.C.: Air Force History and Museum Program, 1999.

Hambling, David, "Military Tackles Problem and Potential of Drones," *Aviation Week and Space Technology*, April 23–May 6, 2018, pp. 44–47.

Hamilton, John, "Air Defense Weapons That Almost Were," *Air Defense Artillery*, July–September 2005, pp. 20–22.

Hamilton, John A., *Blazing Skies: Air Defense Artillery on Fort Bliss, Texas, 1940–2009*, Fort Bliss, Tex.: U.S. Army Air Defense Artillery Center, 2009.

Hammer, John G., and W. R. Elswick, *Conventional Missile Attacks Against Aircraft on Airfields and Aircraft Carriers*, Santa Monica, Calif.: RAND Corporation, RM-4718-PR, 1966. As of May 13, 2020:
https://www.rand.org/pubs/research_memoranda/RM4718.html

Hammer, John G., and Charles A. Sandoval, *Comparison and Evaluation of Protective Alert Shelters for SAC Aircraft*, Santa Monica, Calif.: RAND Corporation, D-8740, 1961. As of May 13, 2020:
https://www.rand.org/pubs/documents/D8740.html

Hammes, T. X. "The Future of Warfare: Small, Many, Smart vs. Few & Exquisite?" War on the Rocks website, July 16, 2014. As of March 9, 2020:
https://warontherocks.com/2014/07/the-future-of-warfare-small-many-smart-vs-few-exquisite/

Hammes, T. X., "In an Era of Cheap Drones, US Can't Afford Exquisite Weapons," Defense One website, January 19, 2016a. As of May 25, 2019:
https://www.defenseone.com/ideas/2016/01/cheap-drones-exquisite-weapons/125216/

Hammes, T. X., "Technologies Converge and Power Diffuses: The Evolution of Small, Smart, and Cheap Weapons," Cato Institute website, January 27, 2016b. As of March 9, 2020:
https://www.cato.org/publications/policy-analysis/technologies-converge-power-diffuses-evolution-small-smart-cheap

Hammes, T. X., "Cheap Technology will Challenge U.S. Tactical Dominance," *Joint Force Quarterly*, Vol. 81, March 29, 2016c. As of March 9, 2020:
https://ndupress.ndu.edu/JFQ/Joint-Force-Quarterly-81/Article/702039/cheap-technology-will-challenge-us-tactical-dominance/

Hampton, Dan, *Viper Pilot: A Memoir of Air Combat*, New York: William Morrow, 2012.

Heavey, Susan, "U.S. Army to Buy Some Israeli-Designed Iron Dome Missile Defenses," Reuters, February 6, 2019. As of July 17, 2019:
https://www.reuters.com/article/us-usa-defense/u-s-army-to-buy-some-israeli-designed-iron-dome-missile-defenses-idUSKCN1PV1QD

Heginbotham, Eric, Michael Nixon, Forrest E. Morgan, Jacob L. Heim, Jeff Hagen, Sheng Li, Jeffrey Engstrom, Martin C. Libicki, Paul DeLuca, David A. Shlapak, David R. Frelinger, Burgess Laird, Kyle Brady, and Lyle J. Morris, *The U.S.-China Scorecard: Forces,*

Geography, and the Evolving Balance of Power, 1996–2017, Santa Monica, Calif.: RAND Corporation, RR-392-AF, 2015. As of March 11, 2020:
https://www.rand.org/pubs/research_reports/RR392.html

Heim, Jacob L., "The Iranian Missile Threat to Air Bases: A Distant Second to China's Conventional Deterrent," *Air and Space Power Journal*, Vol. 29, No. 4, July–August 2015, pp. 27–50.

Hennigan, W. J., "Experts Say Drones Pose a National Security Threat—and We Aren't Ready," *Time*, May 31, 2018. As of May 25, 2019:
http://time.com/5295586/drones-threat/

HESCO, "Mil Units: Protecting Forces Since 1991," webpage, undated. As of July 3, 2019:
https://www.hesco.com/products/mil-units/

Hiatt, Fred, "Improved Missile Too Costly for Pentagon," *Washington Post*, August 28, 1984.

Hill, Kashmir, "Jamming GPS Signals Is Illegal, Dangerous, Cheap and Easy," Gizmodo website, July 24, 2017. As of June 28, 2019:
https://gizmodo.com/jamming-gps-signals-is-illegal-dangerous-cheap-and-e-1796778955

Hinds, Russell A., *The Avenger and SGT York: An Examination of Two Air Defense Systems Non-developmental Item Acquisition Programs*, thesis, Monterey, Calif.: Naval Postgraduate School, March 1995. As of March 9, 2020:
https://apps.dtic.mil/dtic/tr/fulltext/u2/a297415.pdf

Hogg, Ian V., *Anti-Aircraft: A History of Air Defence*, London, UK: MacDonald and Jane's, 1978.

Hogg, Ian V., *Anti-Aircraft Artillery*, Marlborough, UK: Crowood Press, 2002.

Holmes, James M., *The Counterair Companion: A Short Guide to Air Superiority for Joint Force Commanders*, Maxwell AFB, Ala.: Air University Press, 1995.

Holmes, Robert H., "Validating the Abrogation of Joint Service Agreement 8," staff summary sheet, Washington, D.C.: Department of the Air Force, November 18, 2004.

Huie, William Bradford, *Can Do! The Story of the Seabees*, Annapolis, Md.: Naval Institute Press, 1997.

IHS Markit, *Jane's World Air Forces*, 2012.

INFLATECH, "Inflatable Military Decoys," website, undated. As of July 6, 2019:
http://www.inflatechdecoy.com/#pgc-129-1-0

Insinna, Valerie, "Small Drones Still Posing Big Problem for U.S. Air Force Bases," Defense News website, July 14, 2017. As of May 28, 2019:

https://www.defensenews.com/air/2017/07/14/small-drones-still-posing-big-problem-for-us-air-force-bases/

Jackson, Brian A., David R. Frelinger, Michael J. Lostumbo, and Robert W. Button, *Evaluating Novel Threats to the Homeland: Unmanned Aerial Vehicles and Cruise Missiles*, Santa Monica, Calif.: RAND Corporation, MG-626-DTRA, 2008. As of March 11, 2020: https://www.rand.org/pubs/monographs/MG626.html

Jaeger, B. F., and M. B. Schaffer, "Tentative Thoughts on Non-Nuclear IRBM's for Attacking Parked Aircraft," Santa Monica, Calif.: RAND Corporation, D(L)-11285-PR, May 17, 1963. As of May 13, 2020: https://www.rand.org/pubs/documents/D11285.html

Johnson, David E., *Fast Tanks and Heavy Bombers: Innovation in the U.S. Army, 1917–1945*, Ithaca, N.Y.: Cornell University Press, 1998.

Johnson, David E., *Learning Large Lessons: The Evolving Roles of Ground Power and Air Power in the Post-Cold War Era*, Santa Monica, Calif.: RAND Corporation, MG-405-1-AF, 2007. As of March 6, 2020: https://www.rand.org/pubs/monographs/MG405-1.html

Johnson, David E., *Shared Problems: The Lessons of AirLand Battle and the 31 Initiatives for Multi-Domain Battle*, Santa Monica, Calif.: RAND Corporation, PE-301-A/AF, 2018. As of March 6, 2020: https://www.rand.org/pubs/perspectives/PE301.html

Joint Air Power Competence Centre, *Remotely Piloted Aircraft Systems in Contested Environments: A Vulnerability Analysis*, Kalkar, Germany, September 2014. As of March 11, 2020: https://www.japcc.org/portfolio/remotely-piloted-aircraft-systems-in-contested-environments-a-vulnerability-analysis/

Joint Chiefs of Staff Publication No. 8, *Doctrine for Air Defense from Oversea Land Areas*, date unknown.

Joint Publication 3-01, *Countering Air and Missile Threats*, Washington, D.C.: Joint Staff, April 21, 2017, validated May 2, 2018. As of March 11, 2020: https://www.jcs.mil/Portals/36/Documents/Doctrine/pubs/jp3_01_pa.pdf

Joint Publication 3-10, *Doctrine for Rear Area Operations*, Washington, D.C.: Joint Staff, 1992.

Joint Publication 3-13.1, *Electronic Warfare*, Washington, D.C.: Joint Staff, February 8, 2012.

Joint Publication 3-30, *Joint Air Operations*, Washington, D.C.: Joint Staff, July 25, 2019. As of March 11, 2020: https://www.jcs.mil/Portals/36/Documents/Doctrine/pubs/jp3_30.pdf

Joint Staff, Office of the Chairman of the Joint Chiefs of Staff, *DOD Dictionary of Military and Associated Terms*, Washington, D.C., January 2020. As of March 9, 2020:
https://www.jcs.mil/Portals/36/Documents/Doctrine/pubs/dictionary.pdf

Jones, Colin, and Alexander Kirss, "Some Modest Proposals for Defense Department Requirements Reform," War on the Rocks website, August 23, 2018. As of December 15, 2019:
https://warontherocks.com/2018/08/some-modest-proposals-for-defense-department-requirements-reform/

Jones, R. V., *Most Secret War*, London: Wordsworth Editions, 1978.

JP—*See* Joint Publication.

Judson, Jen, "U.S. Army to Prioritize Long-Range Missile Capability to Go After Maritime Targets," Defense News website, March 26, 2019a. As of August 14, 2019:
https://www.defensenews.com/digital-show-dailies/global-force-symposium/2019/03/26/army-to-prioritize-long-range-missile-capability-to-go-after-maritime-targets/

Judson, Jen, "It's Official: U.S. Army Inks Iron Dome Deal," Defense News website, August 12, 2019b. As of August 13, 2019:
https://www.defensenews.com/digital-show-dailies/smd/2019/08/12/its-official-us-army-inks-iron-dome-deal/

Judson, Jen, "U.S. Missile Defense Agency Boss Reveals His Goals, Challenges on the Job," Defense News website, August 19, 2019c. As of August 20, 2019:
https://www.defensenews.com/pentagon/2019/08/19/us-missile-defense-agency-boss-reveals-his-goals-challenges-on-the-job/

Jumper, John P., and Peter J. Schoomaker, "Abrogation of Joint Service Agreement 8," memorandum of understanding, Washington, D.C.: Department of the Air Force and Department of the Army, March 24, 2005.

Karako, Thomas, "Reenergizing the Missile Defense Enterprise," interview with Michael Griffin, Center for Strategic and International Studies website, December 11, 2018. As of March 10, 2020:
https://www.csis.org/analysis/reenergizing-missile-defense-enterprise

Karako, Thomas, "The Missile Defense Review: Insufficient for Complex and Integrated Attack," *Strategic Studies Quarterly*, Summer 2019, pp. 3–15. As of March 9, 2020:
https://www.airuniversity.af.edu/Portals/10/SSQ/documents/Volume-13_Issue-2/Karako.pdf

Keaney, Thomas A., and Eliot A. Cohen, *Gulf War Air Power Survey Summary Report*, Washington, D.C.: U.S. Department of the Air Force, 1993.

King, Scott, and Dennis B. Boykin IV, "Distinctly Different Doctrine: Why Multi-Domain Operations Isn't Airland Battle 2.0," Association of the United States Army website, February 20, 2019. As of July 29, 2019:
https://www.ausa.org/articles/distinctly-different-doctrine-why-multi-domain-operations-isn%E2%80%99t-airland-battle-20

Klimas, Joshua E., *Balancing Consensus, Consent, and Competence: Richard Russell, The Senate Armed Services Committee and Oversight of America's Defense, 1955–1968*, dissertation, The Ohio State University, 2007. As of March 9, 2020:
https://etd.ohiolink.edu/!etd.send_file?accession=osu1196275808&disposition=inline

Koropey, O. B., *It Seemed Like a Great Idea at the Time: The Story of the Sergeant York Air Defense Gun*, Redstone Arsenal, Ala.: Historical Office, U.S. Army Materiel Command, January 1, 1993.

Kozlov, Dmitry, and Sergei Grits, "Russia Says Drone Attacks on Its Bases in Syria are Increasing," Associated Press, August 17, 2018. As of March 11, 2020:
https://apnews.com/2b07cc798d614d84a32ff83f6abe2e7e/Russia-says-drone-attacks-on-its-Syria-base-have-increased

Kreis, John, *Air Warfare and Air Base Air Defense: 1914–1973*, Washington, D.C.: Office of Air Force History, 1988. As of March 9, 2020:
https://apps.dtic.mil/dtic/tr/fulltext/u2/a208631.pdf

Kroesen, Frederick J., "From Cheyenne to Comanche," *Army*, Vol. 55, No. 5, May 2005.

Kube, Courtney, "Russia Has Figured Out How to Jam U.S. Drones in Syria, Officials Say," NBC News website, April 10, 2018. As of June 28, 2019:
https://www.nbcnews.com/news/military/russia-has-figured-out-how-jam-u-s-drones-syria-n863931

Lachow, Irving, "The GPS Dilemma: Balancing Military Risks and Economic Benefits," *International Security*, Vol. 20, No. 1, Summer 1995, pp. 126–148. As of March 11, 2020:
https://www.jstor.org/stable/2539220?seq=1#metadata_info_tab_contents

LaGrone, Sam, "U.S. Navy Allowed to Use Persian Gulf Laser for Defense," USNI News website, December 10, 2014. As of March 10, 2020:
https://news.usni.org/2014/12/10/u-s-navy-allowed-use-persian-gulf-laser-defense

LaGrone, Sam, "Raytheon to Arm Marine Corps with Anti-Ship Missiles in $47M Deal," USNI News website, May 8, 2019. As of August 14, 2019:
https://news.usni.org/2019/05/08/raytheon-to-arm-marine-corps-with-anti-ship-missiles-in-47m-deal

Lambeth, Benjamin S., *The Winning of Air Supremacy in Operation Desert Storm*, Santa Monica, Calif.: RAND Corporation, P-7837, 1993. As of March 11, 2020:
https://www.rand.org/pubs/papers/P7837.html

Lambeth, Benjamin S., *NATO's Air War for Kosovo: A Strategic and Operational Assessment*, Santa Monica, Calif.: RAND Corporation, MR-1365-AF, 2001. As of March 11, 2020:
https://www.rand.org/pubs/monograph_reports/MR1365.html

Lamothe, Dan, "Veil of Secrecy Lifted on Pentagon Office Planning 'Avatar' Fighters and Drone Swarms," *Washington Post*, March 8, 2016.

Lacher, Andrew, Jonathan Baron, Jonathan Rotner, and Michael Balazs, *Small Unmanned Aircraft: Characterizing the Threat*, McLean, Va.: MITRE Corporation, February 2019. As of March 11, 2020:
https://www.mitre.org/publications/technical-papers/small-unmanned-aircraft-characterizing-the-threat

Lane, Jarrett, and Michelle Johnson, "Failures of Imagination: The Military's Biggest Acquisition Challenge," War on the Rocks website, April 3, 2018. As of December 15, 2019:
https://warontherocks.com/2018/04/failures-of-imagination-the-militarys-biggest-acquisition-challenge/

Lappin, Yaakov, "Rafael Launches Spike NLOS from Tomcar Buggy," *Jane's Defence Weekly*, February 6, 2019. As of March 9, 2020:
https://janes.ihs.com/DefenceWeekly/Display/FG_1617467-JDW

Lee, Richard R., *7AF Local Base Defense Operations, July 1965–December 1968*, Hickam AFB, Hawaii: Headquarters Pacific Air Forces, July 1, 1969.

LeMay Center for Doctrine Development and Education, "Doctrine Advisory: Control of the Air," Maxwell AFB, Ala.: Air University, July 31,2017. As of March 12, 2020:
https://www.doctrine.af.mil/Portals/61/documents/doctrine_updates/du_17_01.pdf?ver=2017-09-17-113839-373

Lewis, Donald E., Bruce W. Don, Robert M. Paulson, and Willis W. Ware, *A Perspective on the USAFE Collocated Operating Base System*, Santa Monica, Calif.: RAND Corporation, N-2366-AF, 1986. As of March 6, 2020:
https://www.rand.org/pubs/notes/N2366.html

Lin, Bonny, and Cristina L. Garafola, *Training the People's Liberation Army Air Force Surface-to-Air Missile (SAM) Forces*, Santa Monica, Calif.: RAND Corporation, RR-1414-AF, 2016. As of March 6, 2020:
https://www.rand.org/pubs/research_reports/RR1414.html

Liptak, Andrew, "The U.S. Army Will Test a New GPS That's Resistant to Jamming This Fall," *The Verge* website, June 9, 2019a. As of June 25, 2019:
https://www.theverge.com/2019/6/9/18658901/us-army-testing-jamming-resistant-gps-global-positioning-system-russia-2nd-cavalry-regiment

Liptak, Andrew, "The U.S. Air Force Has a New Weapon Called THOR That Can Take Out Swarms of Drones," *The Verge* website, June 21, 2019b. As of June 24, 2019:
https://www.theverge.com/2019/6/21/18701267/us-air-force-thor-new-weapon-drone-swarms

Lonnquest, John C., and David F. Winkler, *To Defend and Deter: The Legacy of the United States Cold War Missile Program*, Rock Island, Ill.: Defense Publishing Service, U.S. Army Construction Engineering Research Laboratories (USACERL) Special Report 97/01, November 1996. As of March 9, 2020:
https://apps.dtic.mil/dtic/tr/fulltext/u2/a337549.pdf

Lostumbo, Michael J., David R. Frelinger, James Williams, and Barry Wilson, *Air Defense Options for Taiwan: An Assessment of Relative Costs and Operational Benefits*, Santa Monica, Calif.: RAND Corporation, RR-1051-OSD, 2016. As of March 6, 2020:
https://www.rand.org/pubs/research_reports/RR1051.html

Macdonald, Julia, "The Most Surprising Thing About the Venezuela Drone Attack Is That It Hasn't Happened Sooner," *Political Violence at a Glance* website, September 4, 2018. As of March 11, 2020:
https://politicalviolenceataglance.org/2018/09/04/the-most-surprising-thing-about-the-venezuela-drone-attack-is-that-it-hasnt-happened-sooner/

Macias, Amanda, "Russia and China are 'Aggressively Developing' Hypersonic Weapons," CNBC, March 21, 2018. As of March 11, 2020:
https://www.cnbc.com/2018/03/21/hypersonic-weapons-what-they-are-and-why-us-cant-defend-against-them.html

Madrigal, Alexis C., "Drone Swarms Are Going to Be Terrifying and Hard to Stop," *The Atlantic*, March 7, 2018.

Mahnken, Thomas G., *The Cruise Missile Challenge*, Washington, D.C.: Center for Strategic and Budgetary Assessments, 2005. As of March 11, 2020:
https://csbaonline.org/research/publications/the-cruise-missile-challenge

Marinaccio, Richard E., and Ward M. Meier, "Proximity Fuze Jammer," U.S. Patent Number US4121214A, December 16, 1969. As of June 12, 2019:
https://patents.google.com/patent/US4121214A/en

Martini, John A., and Stephen A. Haller, *What We Have, We Shall Defend: An Interim History and Preservation Plan for Nike Site SF-88L, Fort Barry, California*: Part I, San Francisco, Calif.: National Park Service, 1998. As of April 24, 2020:
https://www.nps.gov/goga/learn/historyculture/upload/What%20We%20Have.pdf

McCullough, Amy, "The Looming Swarm," *Air Force Magazine*, April 2019. As of March 9, 2020:
https://www.airforcemag.com/article/the-looming-swarm/

McGarvey, Patrick J., ed., *Visions of Victory: Selected Vietnamese Communist Military Writings, 1964–1968*, Stanford, Calif.: Hoover Institution on War, Revolution, and Peace, 1969.

McIntire, Randall, "The Return of Army Short-Range Air Defense in a Changing Environment," *Fires*, November–December 2017, pp. 5–8. As of March 9, 2020:
https://sill-www.army.mil/firesbulletin/archives/2017/nov-dec/articles/1_McIntire.pdf

McLeary, Paul, "Marines Develop Laser to Fry Drones from JLTVs," Breaking Defense website, June 26, 2019a. As of July 2, 2019:
https://breakingdefense.com/2019/06/marines-develop-laser-to-fry-drones-from-jltvs/

McLeary, Paul, "U.S. Army Signals Israel's Iron Dome Isn't the Answer," Breaking Defense website, October 15, 2019b. As of December 14, 2019:
https://breakingdefense.com/2019/10/us-army-signals-israels-iron-dome-isnt-the-answer/

McMullen, Richard F., *History of Air Defense Weapons: 1946–1962*, Historical Division, Headquarters Air Defense Command, 1963.

McNabb, Harry, "Invisible Interdiction: Air Force Awards Contract for Rail-Mounted Anti-Drone System," Drone Life website, June 12, 2019. As of March 11, 2020:
https://dronelife.com/2019/06/12/invisible-interdiction-air-force-awards-contract-for-rail-mounted-anti-drone-system/

McNaugher, Thomas L., *New Weapons, Old Politics: America's Military Procurement Muddle*, Washington, D.C.: Brookings Institution, 1989.

McPeak, Merrill A., *Roles and Missions*, Oswego, Oreg.: Lost Wingman Press, 2017.

Melyan, Wesley R. C., *The War in Vietnam: 1965*, Project CHECO Report, Honolulu, Hawaii: Headquarters Pacific Air Forces, CHECO Division, January 25, 1967.

Michel, Arthur Holland, *Counter-Drone Systems*, Annandale-on-Hudson, N.Y.: Center for the Study of the Drone, February 2018.

Miller, Thomas G., Jr., *The Cactus Air Force*, New York: Bantam Books, 1987.

Milner, Joseph A., "The Defense of Joint Base Balad: An Analysis," in Caudill, 2014, pp. 217–242.

Missile Defense Advocacy Alliance, "National Advanced Surface-to-Air Missile System (NASAMS)," webpage, undated. As of March 10, 2020:
https://missiledefenseadvocacy.org/defense-systems/national-advanced-surface-to-air-missile-system-nasams/

Missile Defense Project, "Missiles of the World," Missile Threat website, Center for Strategic and International Studies, undated. As of April 22, 2019:
https://missilethreat.csis.org/missile/

Missile Defense Project, "SS-26 Iskander," Missile Threat website, Center for Strategic and International Studies, December 19, 2019. As of March 6, 2020:
https://missilethreat.csis.org/missile/ss-26-2/

Mitchell, William, *Our Air Force: The Keystone of National Defense*, New York: E. P. Dutton and Company, 1921.

Moody, Walton S., and Jacob Neufeld, "Modernizing After Vietnam," in Bernard C. Nalty, ed., *Winged Shield, Winged Sword: A History of the United States Air Force*, Vol. II: *1950–1997*, Washington, D.C.: U.S. Air Force, 1997, pp. 339–372.

Morgan, Mark L., and Mark A. Berhow, *Rings of Supersonic Steel: Air Defenses of the United States Army, 1950–1979*, 3rd ed., Bodega Bay, Calif.: Hole in the Head Press, 2010.

National Materials Advisory Board, *Opportunities in Protective Material Science and Technology for Future Army Applications*, Washington, D.C.: National Academies Press, 2011.

National Photographic Interpretation Center, "Inflatable Dummy Aircraft: Arkhanelsk/Talagi Airfield, USSR," Washington, D.C.: Central Intelligence Agency, June 16, 1983. As of July 6, 2019:
https://www.cia.gov/library/readingroom/print/2013121

Nasser, A., Fathy M. Ahmed, K. H. Moustafa, and Ayman Elshabrawy, "Recent Advancements in Proximity Fuzes Technology," *International Journal of Engineering Research & Technology*, Vol. 4, No. 4, April 2015, pp. 1233–1238.

Naval Air Warfare Center Weapons Division, *Electronic Warfare and Radar Systems Engineering Handbook*, 4th ed., Point Mugu, Calif., 2013. As of March 10, 2020:
https://apps.dtic.mil/dtic/tr/fulltext/u2/a617071.pdf

Osborn, Kris, "New Army Stryker 30mm Cannon Targeting Destroys Moving Drones," Warrior Maven website, April 30, 2019. As of July 2, 2019:
https://defensemaven.io/warriormaven/land/new-army-stryker-30mm-cannon-targeting-destroys-moving-drones-vfW5VtlNUUWurAwI28GJpg/

Ott, David Ewing, *Vietnam Studies: Field Artillery, 1954–1973*, 1995, Washington, D.C.: Department of the Army, 1975. As of March 9, 2020:
https://history.army.mil/html/books/090/90-12/CMH_Pub_90-12.pdf

Pace, Scott, Gerald Frost, Irving Lachow, David Frelinger, Donna Fossum, Donald K. Wassem, and Monica Pinto, *The Global Positioning System: Assessing National Policies*, Santa Monica, Calif.: RAND Corporation, MR-614-OSTP, 1995. As of March 6, 2020:
https://www.rand.org/pubs/monograph_reports/MR614.html

Pacific Air Forces, *Follow-Up to Bien Hoa Mortar Attack*, Project CHECO staff report, Hickam AFB, Hawaii: Headquarters Pacific Air Forces, December 1965a.

Pacific Air Forces, "The Role of Aerospace Security Forces in Limited War Operations," memorandum to Headquarters, U.S. Air Force, December 1, 1965b.

Pacific Air Forces, "Gunship Program for Air Base Defense," Maxwell AFB, Ala.: U.S. Air Force Archives, April 7, 1969.

"Patriot Missile Long-Range Air-Defence System," Army Technology website, undated. As of July 29, 2019:
https://www.army-technology.com/projects/patriot/

Pellerin, Cheryl, "DoD Strategic Capabilities Office Is Near-Term Part of Third Offset," U.S. Department of Defense website, November 3, 2016. As of August 16, 2019:
https://www.defense.gov/Newsroom/News/Article/Article/995438/dod-strategic-capabilities-office-is-near-term-part-of-third-offset/

Phillips, Russell, "An Ineffective System: The M247 Sergeant York," Russell Phillips Military Technology and History website, undated. As of March 12, 2020:
https://russellphillips.uk/an-ineffective-system-the-m247-sergeant-york/

Pinter, William E., *Concentrating on Dispersed Operations: Answering the Emerging Antiaccess Challenge in the Pacific Rim*, thesis, Air University, School of Advanced Air and Space Studies, Maxwell AFB, Ala.: Air University Press, 2007. As of March 9, 2020:
https://apps.dtic.mil/dtic/tr/fulltext/u2/a488573.pdf

Price, Alfred, *War in the Fourth Dimension: U.S. Electronic Warfare from the Vietnam War to the Present*, Mechanicsburg, Pa.: Stackpole Books, 2001.

Price, Alfred, *Instruments of Darkness: The History of Electronic Warfare, 1939–1945*, Yorkshire, UK: Frontline Books, 2017.

Priebe, Miranda, Alan J. Vick, Jacob L. Heim, and Meagan L. Smith, *Distributed Operations in a Contested Environment: Implications for USAF Force Presentation*, Santa Monica, Calif.: RAND Corporation, RR-2959-AF, 2019. As of March 6, 2020:
https://www.rand.org/pubs/research_reports/RR2959.html

Public Law 103-160, National Defense Authorization Act for Fiscal Year 1994, November 30, 1993.

Public Law 115-232, The John S. McCain National Defense Authorization Act for Fiscal Year 2019, August 13, 2018.

Public Law 253, National Security Act of 1947, July 26, 1947.

R., Manfred, *The Conventional Arms Balance*, Part 3: *Deterring Nuclear War in Europe*, Washington, D.C.: The Heritage Foundation, July 16, 1986.

RAND Cost Analysis Section, *The Cost of Decreasing Vulnerability of Air Bases by Dispersal: Dispersing a B-36 Wing*, Santa Monica, Calif.: RAND Corporation, R-235, 1952. As of May 13, 2020:
https://www.rand.org/pubs/reports/R235.html

Rausch, John, "Joint Capability Integration & Development System Overview: New Manual and Sustainment Key Performance Parameters," briefing slides, Washington, D.C.: Joint Staff, August 31, 2018. As of December 15, 2019:
https://www.acq.osd.mil/log/MR/.PSM_workshop.html/2019_Files/Day_One/3_New_JCIDS_Instruction_Guidebook_Rausch.pdf

Raytheon Company, "National Advanced Surface-to-Air Missile System," webpage, undated. As of August 11, 2019:
https://www.raytheon.com/capabilities/products/nasams

Rearden, Steven L., *History of the Office of the Secretary of Defense*, Vol. I: *The Formative Years 1947–1950*, Washington, D.C.: Historical Office, Office of the Secretary of Defense, 1984.

Rector, William, *The Role and Mission of Air Base Defense in a Counterinsurgency War*, Maxwell AFB, Ala.: Project Corona Harvest, Aerospace Studies Institute Air University, May 1970.

Rehberg, Carl, and Mark Gunzinger, *Air and Missile Defense at a Crossroads: New Concepts and Technologies to Defend America's Overseas Bases*, Washington, D.C.: Center for Strategic and Budgetary Assessments, 2018.

Reid, David, "A Swarm of Armed Drones Attacked a Russian Military Base in Syria," CNBC, January 11, 2018. As of March 11, 2020:
https://www.cnbc.com/2018/01/11/swarm-of-armed-diy-drones-attacks-russian-military-base-in-syria.html

Reit, Seymour, *Masquerade: The Amazing Camouflage Deceptions of World War II*, New York: Hawthorn Books, Inc., 1978.

Rempfer, Kyle, "This Gun Shoots Drones out of the Sky," Defense News website, April 10, 2018. As of March 11, 2020:
https://www.defensenews.com/digital-show-dailies/navy-league/2018/04/10/this-gun-shoots-drones-out-of-the-sky/

Richardson, Doug, *An Illustrated Guide to the Techniques and Equipment of Electronic Warfare*, New York: Arco Publishing, 1985.

Robershotte, Mark A., and Greg H. Parlier, "Army Retains Patriot," *Air Defense Artillery*, Spring 1986, pp. 17–19.

Rocco, Domenic P., Jr., "Air Base Defense," *Air Defense Artillery*, Spring 1984, pp. 24–28.

Romjue, John L., *From Active Defense to AirLand Battle: The Development of Army Doctrine 1973–1982*, Fort Monroe, Va.: Historical Office, U.S. Army Training and Doctrine Command, June 1984.

Rubb Military Buildings, "Aircraft Hangars," webpage, 2016. As of July 5, 2019:
http://www.rubbmilitary.com/solutions/aircraft-hangars

Rubin, Alissa J., "Audacious Raid on NATO Base Shows Taliban's Reach," *New York Times*, September 16, 2012.

Rundquist, Erik K., "A Short History of Air Base Defense: From World War I to Iraq," in Caudill, 2014, pp. 3–42.

Rundquist, Eric K., and Raymond J. Fortner, "An Airman Reports: Task Force 455 and the Defense of Bagram Airfield, Afghanistan," in Caudill, 2019, pp. 111–135.

Russell, Stuart, Anthony Aguirre, Ariel Conn, and Max Tegmark, "Why You Should Fear 'Slaughterbots'"—A Response," *IEEE Spectrum*, January 23, 2018. As of March 11, 2020:
https://spectrum.ieee.org/automaton/robotics/artificial-intelligence/why-you-should-fear-slaughterbots-a-response

Scales, Robert H., *Certain Victory: The U.S. Army in the Gulf War*, Sterling, Va.: Potomac Books, Inc., 1993.

Scharre, Paul, *Robotics on the Battlefield*, Part II: *The Coming Swarm*, Washington, D.C.: Center for a New American Security, 2014. As of March 11, 2020:
https://www.cnas.org/publications/reports/robotics-on-the-battlefield-part-ii-the-coming-swarm

Scharre, Paul, "Why You Shouldn't Fear 'Slaughterbots,'" *IEEE Spectrum*, December 22, 2017. As of March 11, 2020:
https://spectrum.ieee.org/automaton/robotics/military-robots/why-you-shouldnt-fear-slaughterbots

Schmitt, Eric, "New Report on Long-Sought Goal: Efficiency in Military," *New York Times*, May 25, 1995.

Schubert, Frank N., and Theresa L. Kraus, *The Whirlwind War: The United States Army in Operations Desert Shield and Desert Storm*, Washington, D.C.: U.S. Government Printing Office, 1995.

Sheftick, Gary, "FY20 Budget to Boost Air and Missile Defense," U.S. Army website, March 13, 2019. As of July 17, 2019:
https://www.army.mil/article/218504/fy20_budget_to_boost_air_missile_defense

Sherman, Jason, "Army Picks Iron Dome for Interim CMD, Eyes Long-Term Adoption of Israeli System," Inside Defense website, January 9, 2019a.

Sherman, Jason, "Army Eyes Laser Weapon for Short-Range Air and Missile Defense by 2027," Inside Defense website, January 10, 2019b.

Sherman, Jason, "MDA Forwards Proposal for New Hypersonic Defense Program to Pentagon for Review," Inside Defense website, August 7, 2019c.

Sherman, Jason, "DOD Readies New Weapon System Plan for 'Regional' Hypersonic Defense," Inside Defense website, December 11, 2019d. As of December 14, 2019:
https://insidedefense.com/daily-news/dod-readies-new-weapon-system-plan-regional-hypersonic-defense

Shlapak, David A., and Alan Vick, *"Check Six Begins on the Ground": Responding to the Evolving Ground Threat to U.S. Air Force Bases*, Santa Monica, Calif.: RAND Corporation, MR-606-AF, 1995. As of March 6, 2020:
https://www.rand.org/pubs/monograph_reports/MR606.html

Shlapak, David A., David T. Orletsky, Toy I. Reid, Murray Scot Tanner, and Barry Wilson, *A Question of Balance: Political Context and Military Aspects of the China-Taiwan Dispute*, Santa Monica, Calif.: RAND Corporation, MG-888-SRF, 2009. As of March 6, 2020:
https://www.rand.org/pubs/monographs/MG888.html

Shugart, Thomas, and Javier Gonzales, *First Strike: China's Missile Threat to U.S. Bases in Asia*, Washington, D.C.: Center for a New American Security, June 2017. As of March 11, 2020:
https://s3.amazonaws.com/files.cnas.org/documents/CNASReport-FirstStrike-Final.pdf?mtime=20170626140814

Sidoti, Sal, *Air Base Operability: A Study in Airbase Survivability and Post-Attack Recovery*, 2nd ed., Canberra: Aerospace Centre, 2001. As of March 11, 2020:
http://airpower.airforce.gov.au/APDC/media/PDF-Files/Fellowship%20Papers/FELL20-Airbase-Operability-A-Study-in-Airbase-Survivalability-and-Post-Attack-Recovery.pdf

Smoley, John Knute, *Seizing Victory from the Jaws of Deterrence: Preservation and Public Memory of America's Nike Air Defense Missile System*, dissertation, Santa Barbara, Calif.: University of California at Santa Barbara, 2008.

Solomon, Jonathan F., "Maritime Deception and Concealment: Concepts for Defeating Wide-Area Oceanic Surveillance-Reconnaissance-Strike Networks," *Naval War College Review*, Vol. 66, No. 4, Autumn 2013.

Southeast Asia Team, Project CHECO, *Attack Against Tan Son Nhut: Project CHECO Southeast Asia Report*, Hickam AFB, Hawaii: Headquarters, Pacific Air Forces, 1966.

Spears, Will, and Ross Hobbs, "A Bomber for the Navy," Over the Horizon website, April 15, 2019. As of May 8, 2019:
https://othjournal.com/2019/04/15/a-bomber-for-the-navy/

Speier, Richard H., George Nacouzi, Carrie A. Lee, and Richard M. Moore, *Hypersonic Missile Nonproliferation: Hindering the Spread of a New Class of Weapons*, Santa Monica, Calif.: RAND Corporation, RR-2137-CC, 2017. As of March 11, 2020:
https://www.rand.org/pubs/research_reports/RR2137.html

Squeo, Anne Marie, "The Assault on Iraq: U.S. Bombs Iraqi GPS-Jamming Sites," *Wall Street Journal*, March 26, 2003. As of June 28, 2019:
https://search.proquest.com/docview/398805878/AFDB5933C7044F97PQ/32?accountid=25%20333

Starnes, E. C., "U.S. Roland Now Totally Fielded with 'Total Army,'" *Air Defense Artillery*, July–August 1987, pp. 49–50.

Starr, Barbara, "US General Warns of Hypersonic Weapons Threat from Russia and China," CNN, March 27, 2018. As of March 11, 2020:
https://www.cnn.com/2018/03/27/politics/general-hyten-hypersonic-weapon-threat/index.html

Stewart, Richard W., ed., *American Military History*, Vol. II, *The United States Army in a Global Era, 1917–2008*, Washington, D.C.: Center of Military History, United States Army, 2010.

Stillion, John, *Trends in Air-to-Air Combat: Implications for Future Air Superiority*, Washington, D.C.: Center for Strategic and Budgetary Assessments, 2015.

Stillion, John, and David T. Orletsky, *Airbase Vulnerability to Conventional Cruise-Missile and Ballistic-Missile Attacks: Technology, Scenarios, and U.S. Air Force Responses*, Santa Monica, Calif.: RAND Corporation, MR-1028-AF, 1999. As of March 6, 2020:
https://www.rand.org/pubs/monograph_reports/MR1028.html

Stillion, John, and Bryan Clark, *What It Takes to Win: Succeeding in 21st Century Battle Network Competitions*, Washington, D.C.: Center for Strategic and Budgetary Assessments, 2015.

Stuart, Douglas T., ed., *Organizing for National Security*, Carlisle, Pa.: Army War College Strategic Studies Institute, 2000.

SachuHopes [Shaheed Brahim], user's own photograph of DJI phantom drone, Wikimedia Commons website, licensed under the Creative Commons Attribution-Share Alike 4.0 International license, 2018. As of March 11, 2020:
https://commons.wikimedia.org/wiki/File:DJI_phantom_Drone.jpg

Thomas, Brent, Mahyar A. Amouzegar, Rachel Costello, Robert A. Guffey, Andrew Karode, Christopher Lynch, Kristin F. Lynch, Ken Munson, Chad J. R. Ohlandt, Daniel M. Romano, Ricardo Sanchez, Robert S. Tripp, and Joseph V. Vesely, *Project AIR FORCE Modeling Capabilities for Support of Combat Operations in Denied Environments*, Santa Monica, Calif.: RAND Corporation, RR-427-AF, 2015. As of March 6, 2020:
https://www.rand.org/pubs/research_reports/RR427.html

Thompson, George Raynor, Dixie R. Harris, Pauline M. Oakes, and Dulany Terrett, *The Signal Corps: The Test (December 1941 to July 1943)*, U.S. Army in World War II, The Technical Services, Washington, D.C.: Office of the Chief of Military History, Department of the Army, 1957.

Thompson, George Raynor, and Dixie R. Harris, *The Signal Corps: The Outcome (Mid-1943 Through 1945)*, U.S. Army in World War II, The Technical Services, Washington, D.C.: Office of the Chief of Military History, 1966.

Thornhill, Paula, *Demystifying the American Military: Institutions Evolution and Challenges Since 1789*, Annapolis, Md.: Naval Institute Press, 2019.

Torgerson, Frederick, *Parked Aircraft Vulnerability to Mortar Attack*, Hickam AFB, Hawaii: Headquarters Pacific Air Forces, September 1964.

Tressel, Ashley, "Official: Army Two Steps Behind Air and Missile Defense Integration," Inside Defense website, May 7, 2019.

Trest, Warren A., *Air Force Roles and Missions: A History*, Washington, D.C.: Air Force History and Museums Program, 1998. As of March 10, 2020:
https://media.defense.gov/2010/Sep/22/2001330059/-1/-1/0/AFD-100922-020.pdf

Trimble, Steve, "U.S. Air Force Revives Neglected Sea-Strike Mission," *Aviation Week and Space Technology*, April 30, 2019.

Truman, Harry S., *Memoirs by Harry S. Truman*, Vol. 2: *Years of Trial and Hope*, Garden City, N.Y.: Doubleday & Company, Inc., 1956.

Tuck, R. E., *Preservation of Tactical Air Combat Potential in Western Europe: Guided Missile Defense Potential*, Santa Monica, Calif.: RAND Corporation, RM-1312, 1954. As of May 13, 2020:
https://www.rand.org/pubs/research_memoranda/RM1312.html

UAS Task Force, *Unmanned Aircraft System Airspace Integration Plan*, Washington, D.C.: U.S. Department of Defense, March 2011. As of March 11, 2020:
https://info.publicintelligence.net/DoD-UAS-AirspaceIntegration.pdf

USAF—*See* U.S. Air Force.

USAFE—*See* U.S. Air Forces Europe.

U.S. Air Force, AFWERX website, undated. As of August 16, 2019:
https://www.afwerx.af.mil

U.S. Air Force, *Gulf War Air Power Survey*, Vol. III: *Logistics and Support*, Washington, D.C.: Headquarters U.S. Air Force, 1993a.

U.S. Air Force, *Global Vigilance, Global Reach, Global Power for America*, Washington, D.C.: Department of the Air Force, 2013b.

U.S. Air Force, *The World's Greatest Air Force, Powered by Airmen, Fueled by Innovation: A Vision for the United States Air Force*, Washington, D.C.: Department of the Air Force, 2013c. As of March 12, 2020:
https://www.airforcemag.com/PDF/DocumentFile/Documents/2013/vision_011013.pdf

U.S. Air Force, *Global Horizons Final Report United States Air Force Global Science and Technology Vision,* Washington, D.C.: Department of the Air Force, June 21, 2013d. As of March 12, 2020:
https://www.airforcemag.com/PDF/DocumentFile/Documents/2013/
GlobalHorizons_062313.pdf

U.S. Air Force, *America's Air Force: A Call to the Future*, Washington, D.C.: Department of the Air Force, July 2014. As of March 12, 2020:
https://www.af.mil/Portals/1/documents/SECAF/AF_30_Year_Strategy.pdf

U.S. Air Force, *Air Force Future Operating Concept: A View of the Air Force in 2035*, Washington, D.C.: Department of the Air Force, September 2015. As of March 12, 2020:
https://www.af.mil/Portals/1/images/airpower/AFFOC.pdf

U.S. Air Force, Doctrine Annex 3-51, *Electronic Warfare*, Maxwell AFB: Curtis E. LeMay Center for Doctrine Development and Education, October 10, 2014.

U.S. Air Force Inspector General SS, "Manpower Restrictions," memorandum, in *Defense of Air Bases Inspector General Report*, Vol. V, No. 1, February 17, 1968.

U.S. Air Forces Europe, "Protection of the Forces," in U.S. Air Forces Europe, *History of United States Air Forces in Europe: 1 July–31 December 1960*, Vol. 1: *Narrative*, Ramstein AB, Germany: Office of the Command Historian, 1960, p. 58.

U.S. Air Forces Europe, "Structure and Realignment of Air Defense Forces," in U.S. Air Forces Europe, *History of United States Air Forces in Europe: 1 January–30 June 1960*, Vol. 1: *Narrative*, Ramstein AB, Germany: Office of the Command Historian, 1961, pp. 44–67.

U.S. Air Forces Europe, "USAFE Shelter Construction Status," in U.S. Air Forces Europe, *History of United States Air Forces in Europe: For Fiscal Year 1972*, Vol. 1: *Narrative*, Ramstein AB, Germany: Office of the Command Historian, 1973, pp. 56–57.

U.S. Air Forces Europe, *History of United States Air Forces in Europe for Fiscal Year 1972*, Vol. I: *Narrative*, Ramstein AB, Germany: Office of the Command Historian, 1973.

U.S. Air Forces Europe, *History of United States Air Forces in Europe: Calendar Year 1977*, Vol. 1: *Narrative*, Ramstein AB, Germany: Office of the Command Historian, 1978.

U.S. Air Forces Europe, *History of United States Air Forces in Europe: Calendar Year 1983*, Vol. 1: *Narrative*, Ramstein AB, Germany: Office of the Command Historian, 1984.

U.S. Air Forces Europe, *History of United States Air Forces in Europe: Calendar Year 1985*, Vol. 1: *Narrative*, Ramstein AB, Germany: Office of the Command Historian, 1986.

"U.S. Army Selects European Missile," *New York Times*, January 10, 1975.

U.S. Army, "Counter-Rocket Artillery, Mortar (C-RAM) Intercept Land-Based Phalanx Weapon System (LPWS)," webpage, undated. As of August 8, 2019:
https://asc.army.mil/web/portfolio-item/ms-c-ram_lpws/

U.S. Army, *Operational Report of Headquarters, II Field Force Vietnam Artillery for Period Ending 31 July 1968*, August 15, 1968, Department of the Army.

U.S. Army Space and Missile Defense Command, *Army Air and Missile Defense 2028*, Huntsville, Ala.: USASMDC/ARSTRAT, 2019. As of March 9, 2020:
https://www.smdc.army.mil/Portals/38/Documents/Publications/Publications/
SMDC_0120_AMD-BOOK_Finalv2.pdf

U.S. Code, Title 10, Armed Forces, Subtitle A, General Military Law, Pt. I, Organization and General Military Powers, Ch. 7, Boards, Councils, and Committees, Sec. 171, Armed Forces Policy Council.

U.S. Department of Defense, *Conduct of the Persian Gulf War: Final Report to Congress*, Washington, D.C.: April 1992.

U.S. Department of Defense, Defense Innovation Board website, undated. As of August 16, 2019:
https://innovation.defense.gov

U.S. Government Accountability Office, *Missile Defense: Some Progress Delivering Capabilities, but Challenges with Testing Transparency and Requirements Development Need to Be Addressed*, Washington, D.C., GAO-15-381, May 2017.

U.S. House of Representatives, H.R. 3186, U.S.-Israel Indirect Fire Protection Act of 2019, 116th Cong., 2nd Sess., introduced June 10, 2019.

U.S. Military Assistance Command Headquarters, Vietnam, "Request for Security Forces," message to Commanding General, U.S. Army, Vietnam, January 12, 1966.

U.S. Senate, "Study of Airpower," hearings before the Subcommittee of the Air Force of the Committee on Armed Services, Part XVII, 84th Cong., 2nd Sess., June 18 and 25, 1956a.

U.S. Senate, "Study of Airpower," hearings before the Subcommittee of the Air Force of the Committee on Armed Services, Part XVIII, 84th Cong., 2nd Sess., June 18 and 27, 1956b.

U.S. Senate, *Authorizing Construction for Military Departments*, Senate Committee on Armed Services report, 84th Cong., 2nd Sess., July 25, 1956c.

Vigdor, Neil, "What We Know About the 2 Bases Iran Attacked," *New York Times*, January 7, 2020.

Vick, Alan, *Snakes in the Eagle's Nest: A History of Ground Attacks on Air Bases*, Santa Monica, Calif.: RAND Corporation, MR-553-AF, 1995. As of March 6, 2020:
https://www.rand.org/pubs/monograph_reports/MR553.html

Vick, Alan J., *Air Base Attacks and Defensive Counters: Historical Lessons and Future Challenges*, Santa Monica, Calif.: RAND Corporation, RR-968-AF, 2015a. As of March 6, 2020:
https://www.rand.org/pubs/research_reports/RR968.html

Vick, Alan J., *Proclaiming Airpower: Air Force Narratives and American Public Opinion from 1917 to 2014*, Santa Monica, Calif.: RAND Corporation, RR-1044-AF, 2015b. As of March 10, 2020:
https://www.rand.org/pubs/research_reports/RR1044.html

Walker, James, Lewis Bernstein, and Sharon Lang, *Seize the High Ground: The Army in Space and Missile Defense*, Redstone Arsenal, Ala.: Historical Office, U.S. Army Space and Missile Defense Command, 2003.

Wall, Robert, "U.K. Airport Remains Closed After Drones Disrupt Travel," *Wall Street Journal*, December 20, 2018. As of March 11, 2020:

https://www.wsj.com/articles/u-s-bound-flights-from-major-u-k-airport-grounded-over-drone-fears-11545295342

Wallace, Ryan J., and Jon M. Loffi, "Examining Unmanned Aerial System Threats & Defenses: A Conceptual Analysis" *International Journal of Aviation, Aeronautics and Aerospace*, Vol. 2, No. 4, 2015. As of May 22, 2019: https://commons.erau.edu/cgi/viewcontent.cgi?article=1084&context=ijaaa

Warwick, Graham, "DARPA Sees Autonomous Drone Swarms Supporting Urban Operations," *Aviation Week and Space Technology*, March 30, 2018, p. 64. As of March 9, 2020: https://aviationweek.com/defense-space/darpa-sees-autonomous-drone-swarms-supporting-urban-operations

Weitze, Karen, *Eglin Air Force Base: Installation Buildup for Research, Test and Evaluation and Training*, Eglin AFB, Fla.: Air Force Materiel Command, 2001.

Welch, Larry D., D. L. Briggs, R. D. Bleach, G. H. Canavan, S. L. Clark-Sestak, R. W. Constantine, C. W. Cook, M. S. Fries, D. E. Frost, D. R. Graham, D. J. Keane, S. D. Kramer, P. L. Major, C. A. Primmerman, J. M. Ruddy, G. R. Schneiter, J. M. Seng, R. M. Stein, S. D. Weiner, and J. D. Williams, "Study on the Mission, Roles, and Structure of the Missile Defense Agency (MDA)," Alexandria, Va.: Institute for Defense Analyses, IDA Paper P-4374, 2008. As of March 10, 2020: https://apps.dtic.mil/dtic/tr/fulltext/u2/a486276.pdf

Werrell, Kenneth P., *The Evolution of the Cruise Missile*, Maxwell AFB, Ala.: Air University Press, 1985.

Werrell, Kenneth P., *Archie, Flak, AAA, and SAM*, Maxwell AFB, Ala.: Air University Press, 1988. As of March 10, 2020: https://apps.dtic.mil/dtic/tr/fulltext/u2/a421867.pdf

Westermann, Edward B., *Flak: German Anti-Aircraft Defenses, 1941–1945*, Lawrence, Kan.: University Press of Kansas, 2001.

Westmoreland, William C., *A Soldier Reports*, Garden City, N.Y.: Doubleday & Company, 1976.

"What Are Hypersonic Weapons?" *The Economist*, January 3, 2019. As of March 11, 2020: https://www.economist.com/the-economist-explains/2019/01/03/what-are-hypersonic-weapons

White, John, "Preface," in Commission on Roles and Missions of the Armed Forces, 1995. As of March 10, 2020: https://apps.dtic.mil/dtic/tr/fulltext/u2/a295228.pdf

Williams, Theodore C., "US Air Force Ground Defense System," essay, Carlisle Barracks, Pa.: U.S. Army War College, December 2, 1968.

Wilson, Charles E., "Clarification of Roles and Missions to Improve the Effectiveness of Operation of the Department of Defense," memorandum for members of the Armed Forces Policy Council, Washington, D.C., November 26, 1956, as reproduced in Wolf, 1987, pp. 293–301.

Winton, Harold R., "An Ambivalent Partnership: US Army and Air Force Perspectives on Air-Ground Operations, 1973–1990," in Phillip S. Meilinger, ed., *The Paths of Heaven: The Evolution of Airpower Theory*, Maxwell AFB, Ala.: Air University Press, 1997, pp. 399–441. As of March 10, 2020:
https://www.airuniversity.af.edu/Portals/10/AUPress/Books/
B_0029_MEILINGER_PATHS_OF_HEAVEN.pdf

Wolf, Richard I., *The United States Air Force: Basic Documents on Roles and Missions*, Washington, D.C.: Office of Air Force History, 1987. As of March 10, 2020:
https://media.defense.gov/2010/May/25/2001330272/-1/-1/0/AFD-100525-080.pdf

Yanarella, Ernest J., *The Missile Defense Controversy: Technology in Search of a Mission*, Lexington, Ky.: University Press of Kentucky, 2002.

"Yemen's Houthi Rebels Attack Saudi's Najran Airport—Again," Al Jazeera, May 23, 2019. As of March 11, 2020:
https://www.aljazeera.com/news/2019/05/yemen-houthi-rebels-attack-saudi-najran-airport-190523140308211.html

Zaffiri, Samuel, *Westmoreland: A Biography of General William C. Westmoreland*, New York: William Morrow and Company, Inc., 1994.